# STUDENT'S SOLUTIONS MANUAL

## TO ACCOMPANY

# DEVELOPMENTAL MATH

## CANADIAN APPLICATIONS

**PATRICIA PROBERT**

Seneca College

**ANTHONY TARGA**

Seneca College

NELSON EDUCATION

Published in 1997 by
Nelson Education Ltd.

1120 Birchmount Road
Scarborough, Ontario M1K 5G4

Visit our Web site at **http://www.nelson.com**



**Canadian Cataloguing in Publication Data**

Probert, Patricia Joan, date-

Student's solutions manual to accompany Developmental math: Canadian applications

ISBN 0-17-605649-1

1. Mathematics - Problems, exercises, etc.
I. Targa, Anthony, date- . II. Probert, Patricia Joan, date- . Developmental math.
III. Title.

QA39.2.P76 1997 Suppl. 510 C97-930140-8

**ISBN 13: 978-0-17-605649-0**
**ISBN 10: 0-17-605649-1**

Printed and bound in Canada
13 14 15 16 16 15 14 13

# CHAPTER 1: Introduction to Numbers

## Exercise 1.3

2. $2^4 = 2 \times 2 \times 2 \times 2$
$= 16$

4. $\sqrt{64} = +8 \; or \; -8$

since: $(+8) \times (+8) = 64$

and: $(-8) \times (-8) = 64$

6. $8^{-\frac{1}{3}} = \frac{1}{8^{\frac{1}{3}}}$
$= \frac{1}{\sqrt[3]{8}}$
$= \frac{1}{2}$

8. $10^5 = 10 \times 10 \times 10 \times 10 \times 10$
$= 10000$

10. $10^{-8} = \frac{1}{10 \times 10 \times 10 \times 10 \times 10 \times 10 \times 10 \times 10}$
$= \frac{1}{100000000}$

## Exercise 1.4

2. $8 \times 2 + 1 = 16 + 1$
$= 17$

4. $12 \times (5 - 1) + 3 \times 2$
$= 12 \times 4 + 6$
$= 48 + 6$
$= 54$

6. $2 \div (4 - 3)$
$= 2 \div 1$
$= 2$

8. $(4 + 2)^2$
$= 6^2$
$= 36$

10. $2^2 + 3^2$
$= 4 + 9$
$= 13$

## Exercise 1.6

2.a. $24 \times 98 = 24 \times (100 - 2)$
$= 24 \times 100 - 24 \times 2$
$= 2400 - 48$
$= 2352$

b. $5 \times 23 \times 20 = 5 \times 20 \times 23$
$= 100 \times 23$
$= 2300$

c. $23 + 4 + 96 = 23 + 100$
$= 123$

d. $18(20 + 5) = 18 \times 20 + 18 \times 5$
$= 18 \times 10 \times 2 + 9 \times 2 \times 5$
$= 18 \times 2 \times 10 + 9 \times 10$
$= 36 \times 10 + 90$
$= 360 + 90$
$= 450$

## End of Chapter 1 Test

2.a. $|-5| = 5$

because it is 5 units from 0

b. $|10| = 10$

because it is 10 units from 0

c. $|1| = 1$

because it is 1 unit from 0

d. $|-8| = 8$

because it is 8 units from 0

4.a. $2 \times 4 + 3 \times 2 = 8 + 6$

$= 14$

b. $8 - 2 \times 8 - 8$

$= 8 - 8$
$= 0$

c.
$12 + 2 \times (4 - 2) - 4 + 2 \times (8 + 5)$

$= 12 + 2 \times 2 - 4 + 2 \times 13$
$= 12 + 4 - 4 + 26$
$= 38$

d. $2^2 \times 5 = 2 \times 2 \times 5$

$= 4 \times 5$
$= 20$

e. $12 \times 3 + 2^2 = 36 + 2 \times 2$

$= 36 + 4$
$= 40$

f. $(2 - 5)^2 = (-3)^2$

$= (-3) \times (-3)$
$= +9$

# CHAPTER 2: Fractions

Exercise 2.1

2. 3 numerator, 5 denominator

4. 22 numerator, 78 denominator

Exercise 2.2

2. Proper fraction

4. Improper fraction

6. Mixed numeral

Exercise 2.3

2. $\frac{9}{5} = 9 \div 5 = 1\frac{4}{5}$

4. $\frac{33}{10} = 33 \div 10 = 3\ \frac{3}{10}$

6. $\frac{25}{6} = 25 \div 6 = 4\ \frac{1}{6}$

8. $\frac{9}{8} = 9 \div 8 = 1\frac{1}{8}$

10. $\frac{31}{11} = 31 \div 11 = 2\ \frac{9}{11}$

12. $\frac{69}{5} = 69 \div 5 = 13\frac{4}{5}$

14. $\frac{37}{9} = 37 \div 9 = 4\frac{1}{9}$

16. $\frac{333}{100} = 333 \div 100 = 3\frac{33}{100}$

18. $\frac{265}{7} = 265 \div 7 = 37\frac{6}{7}$

20. $3\frac{5}{9} = \frac{32}{9}$ $(3 \times 9) + 5 = 32$

22. $4\frac{5}{6} = \frac{29}{6}$ $(4 \times 6) + 5 = 29$

24. $6\frac{2}{3} = \frac{20}{3}$ $(6 \times 3) + 2 = 20$

26. $10\frac{2}{7} = \frac{72}{7}$ $(10 \times 7) + 2 = 72$

28. $20\ \frac{3}{7} = \frac{143}{7}$ $(20 \times 7) + 3 = 143$

30. $9\frac{5}{8} = \frac{77}{8}$ $(9 \times 8) + 5 = 77$

32. $12\frac{8}{9} = \frac{116}{9}$ $(12 \times 9) + 8 = 116$

34. $5\frac{27}{35} = \frac{202}{35}$ $(5 \times 35) + 27 = 202$

Exercise 2.4

2. 1, 2, 4, 8, 16

4. 1, 3, 5, 9, 15, 45

6. 1, 2, 5, 10

8. 1, 5, 7, 35

10. 1, 2, 3, 4, 6, 8, 12, 24

Exercise 2.5

2. 5, 10, 15, 20, 25, 30, 35, 40, 45, 50

Exercise 2.6

2. 2, 4, <u>6</u>, 8, 10, <u>12</u>, 14, 16, <u>18</u>, ...
   3, <u>6</u>, 9, 12, 15, <u>18</u>, ...
   <u>6, 12, 18</u>, ...

Three common multipliers of 2, 3 and 6 are 6, 12, and 18.

Exercise 2.7

2. prime

4. prime

6. prime

8. not prime, nor composite

Exercise 2.8

2. $\frac{15}{3} = \mathbf{5}$

$15 = 3 \times 5$

4. $\frac{125}{5} = 25 \qquad \frac{25}{5} = \mathbf{5}$

$125 = 5 \times 5 \times 5$

6. $\frac{78}{2} = 39 \qquad \frac{39}{3} = \mathbf{13}$

$78 = 2 \times 3 \times 13$

8. $\frac{46}{2} = \mathbf{23}$

$46 = 2 \times 23$

10. $\frac{169}{13} = 13$

$169 = 13 \times 13$

12. $\frac{98}{2} = 49 \qquad \frac{49}{7} = \mathbf{7}$

$98 = 2 \times 7 \times 7$

Exercise 2.9

2. $5 = 1 \times 5$
$8 = 2 \times 2 \times 2$

$LCM = 2 \times 2 \times 2 \times 5 = 40$

4. $89 = 1 \times 89$
$7 = 1 \times 7$

$LCM = 7 \times 89 = 623$

6. $12 = 2 \times 2 \times 3$
$16 = 2 \times 2 \times 2 \times 2$

$LCM = 2 \times 2 \times 2 \times 2 \times 3 = 48$

8. $3 = 3 \times 1$
$56 = 2 \times 2 \times 2 \times 7$

$LCM = 3 \times 2 \times 2 \times 2 \times 7 = 168$

10. $3 = 1 \times 3$
$12 = 2 \times 2 \times 3$
$18 = 2 \times 3 \times 3$

$LCM = 2 \times 2 \times 3 \times 3 = 36$

12. $2 = 1 \times 2$
$5 = 1 \times 5$
$32 = 2 \times 2 \times 2 \times 2 \times 5 = 160$

$LCM = 2 \times 2 \times 2 \times 2 \times 2 \times 5 = 160$

Exercise 2.10

2. $\frac{1}{8} \times \frac{7}{5} = \frac{1 \times 7}{8 \times 5} = \frac{7}{40}$

4. $7 \times \frac{4}{5} = \frac{7}{1} \times \frac{4}{5} = \frac{7 \times 4}{1 \times 5} = \frac{28}{5} = 5\frac{3}{5}$

6. $\frac{1}{8} \times \frac{5}{7} = \frac{1 \times 5}{8 \times 7} = \frac{5}{56}$

8. $\frac{3}{4} \times \frac{3}{8} = \frac{3 \times 3}{4 \times 8} = \frac{9}{32}$

10. $\frac{12}{8} \times \frac{6}{5} = \frac{12 \times 6}{8 \times 5} = \frac{72}{40} = \frac{8}{8} \times \frac{9}{5} = \frac{9}{5} = 1\frac{4}{5}$

12. $\frac{11}{8} \times \frac{6}{4} = \frac{11 \times 6}{8 \times 4} = \frac{66}{32} = \frac{2}{2} \times \frac{33}{16} = \frac{33}{16} = 2\frac{1}{16}$

14. $\frac{11}{23} \times \frac{7}{5}=\frac{11\times7}{23\times5}=\frac{77}{115}$

16. $3\frac{1}{4} \times 2\frac{1}{2}=\frac{13}{4} \times \frac{5}{2}=\frac{13\times5}{4\times2}=\frac{65}{8}=8\frac{1}{8}$

18. $6\frac{2}{3} \times 9\frac{1}{3}=\frac{20}{3} \times \frac{28}{3}=\frac{560}{9}=62\frac{2}{9}$

20. $2\frac{1}{2} \times 3\frac{2}{5}=\frac{5}{2} \times \frac{17}{5}=\frac{5}{5} \times \frac{17}{2}=\frac{17}{2}=8\frac{1}{2}$

22. $4\frac{5}{8} \times 6\frac{1}{4}=\frac{37}{8} \times \frac{25}{4}=\frac{925}{32}=28\frac{29}{32}$

24. $9\frac{2}{3} \times 3\frac{4}{5}=\frac{29}{3} \times \frac{19}{5}=\frac{551}{15}=36\frac{11}{15}$

26. $2\frac{1}{2} \times 3=\frac{5}{2} \times \frac{3}{1}=\frac{15}{2}=7\frac{1}{2}$
We require $7\frac{1}{2}$ cups of raisins.
28. $33\frac{1}{3} \times 30=\frac{100}{3} \times 30=\frac{100}{3} \times \frac{30}{1}=$

$\frac{3000}{3}=1000$

The record turns 1000 revolutions in 30 minutes of play.

30. $600 \times \frac{1}{2}=\frac{600}{1} \times \frac{1}{2}=\frac{600}{2}=300$

The book has a mass of 300 grams.

32. $\frac{2}{3} \times x = 2400$

$(\frac{3}{2}) \quad \frac{2}{3} \times x = 2400 \quad (\frac{3}{2})$

$\frac{6}{6}x = \frac{2400\times3}{2}$

$x = \frac{7200}{2}$

$x = 3600$

The capacity of the pitcher is 3600 milliliters.

## Exercise 2.11

2. $\frac{3}{4}=\frac{75}{100}$ $\quad 3 \times 100=300$
$4 \times 75=300$

300=300
True

4. $\frac{22}{11}=\frac{2}{1}$ $\quad 22 \times 1=22$
$11 \times 2=22$

22=22
True

6. $\frac{7}{8}=\frac{14}{16}$ $\quad 7 \times 6=112$
$8 \times 14=112$

112=112
True

8. $\frac{5}{12}=\frac{10}{23}$ $\quad 5 \times 23=115$
$12 \times 10=120$

$115 \neq 120$
False

10. $\frac{2}{32}=\frac{1}{16}$ $\quad 2 \times 16=32$
$32 \times 1=32$

32=32
True

12. $\frac{2}{300}=\frac{6}{900}$ $\quad 2 \times 900=1800$
$300 \times 6=1800$

1800=1800
True

Exercise 2.12

2. The reciprocal of $\frac{1}{3}$ is $\frac{3}{1}$, which is 3.

4. The reciprocal of $55\frac{1}{2}$, is $\frac{2}{111}$, since $55\frac{1}{2}$ is $\frac{111}{2}$.

6. The reciprocal of $\frac{69}{5}$ is $\frac{5}{69}$.

Exercise 2.13

2. $\frac{1}{2} \div \frac{5}{9}=\frac{1}{2} \times \frac{9}{5}=\frac{9}{10}$

4. $\frac{23}{25} \div \frac{6}{7}=\frac{23}{25} \times \frac{7}{6}=\frac{161}{150}=1\frac{11}{150}$

6. $\frac{22}{34} \div \frac{11}{13}=\frac{22}{34} \times \frac{13}{11}=\frac{286}{374}=\frac{26}{34}$

8. $6 \div \frac{1}{2}=\frac{6}{1} \times \frac{2}{1}=\frac{12}{1}=12$

10. $\frac{12}{45} \div 3=\frac{12}{45} \times \frac{1}{3}=\frac{12}{135}=\frac{4}{45}$

12. $\frac{9}{13} \div 8=\frac{9}{13} \times \frac{1}{8}=\frac{9}{104}$

14. $3\frac{2}{7} \div 5=\frac{23}{7} \times \frac{1}{5}=\frac{23}{35}$

16. $\frac{1}{2} \div 3\frac{1}{3}=\frac{1}{2} \div \frac{10}{3}=\frac{1}{2} \times \frac{3}{10}=\frac{3}{20}$

18. $\frac{5}{6} \times x = 600$

$\left(\frac{6}{5}\right) \quad \frac{5}{6} \times x = 600 \quad \left(\frac{6}{5}\right)$

$\frac{30}{30} \times x = \frac{600}{1} \times \frac{6}{5}$

$x = \frac{3600}{5}$

$x = 720$

Joshua 's total journey is 720 kilometers.

20. $150 \div \frac{3}{5}$

$= 150 \times \frac{5}{3}$

$= \frac{150}{1} \times \frac{5}{3}$

$= \frac{150 \times 5}{1 \times 3}$

$= \frac{750}{3}$

$= 250$

The 150 millimeter vile contains 250 doses of vaccine.

22. $100 \div 3\frac{1}{3}$

$= \frac{100}{1} \div \frac{10}{3}$

$= \frac{100}{1} \times \frac{3}{10}$

$= \frac{100 \times 3}{1 \times 10}$

$= \frac{300}{10}$

$= 30$

30 dresses can be made with the 100 meter bolt of fabric.

Exercise 2.14

2. $\frac{2}{8}+\frac{5}{8}=\frac{7}{8}$

4. $\frac{4}{25}+\frac{6}{25}=\frac{10}{25}=\frac{2}{5}$

6. $\frac{2}{9}+\frac{4}{8}=\frac{16}{72}+\frac{36}{72}=\frac{52}{72}$

8. $\frac{6}{40}+\frac{3}{13}=\frac{78}{520}+\frac{120}{520}=\frac{198}{520}$

10. $\frac{1}{3}+\frac{6}{7}=\frac{7}{21}+\frac{18}{21}=\frac{25}{21}=1\frac{4}{21}$

12. $\frac{1}{13}+\frac{3}{11}=\frac{11}{143}+\frac{39}{143}=\frac{50}{143}$

14. $5\frac{1}{5}+2\frac{2}{5}=\frac{26}{5}+\frac{12}{5}=\frac{38}{5}=7\frac{3}{5}$

16. $6\frac{2}{5}+1\frac{3}{5}=\frac{32}{5}+\frac{8}{5}=\frac{40}{5}=8$

18. $5\frac{1}{5}+4\frac{2}{9}=\frac{26}{5}+\frac{38}{9}=\frac{234}{45}+\frac{190}{45}=\frac{424}{45}=9\frac{19}{45}$

20. $4\frac{2}{7}+3\frac{1}{3}=\frac{30}{7}+\frac{10}{3}=\frac{90}{21}+\frac{70}{21}=\frac{160}{21}=7\frac{13}{21}$

22. $5\frac{1}{16}+2\frac{1}{3}=\frac{81}{16}+\frac{7}{3}=\frac{243}{48}+\frac{112}{48}=\frac{355}{48}=7\frac{19}{48}$

24. $5\frac{6}{7}+3\frac{2}{9}=\frac{41}{7}+\frac{29}{9}=\frac{369}{63}+\frac{203}{63}=\frac{572}{63}=9\frac{5}{63}$

26. $56\frac{1}{3}+123\frac{3}{5}=179\frac{10+18}{30}=179\frac{28}{30}=179\frac{14}{15}$

A total of $179\frac{14}{15}$ meters of fence can be erected with the two rolls of fencing.

Exercise 2.15

2. $\frac{5}{7}-\frac{1}{7}=\frac{4}{7}$

4. $\frac{12}{33}-\frac{5}{33}=\frac{7}{33}$

6. $\frac{9}{13}-\frac{1}{2}=\frac{18}{26}-\frac{13}{26}=\frac{5}{26}$

8. $\frac{8}{9}-\frac{1}{3}=\frac{24}{27}-\frac{9}{27}=\frac{15}{27}=\frac{5}{9}$

10. $\frac{6}{7}-\frac{5}{8}=\frac{48}{56}-\frac{35}{56}=\frac{13}{56}$

12. $\frac{5}{7}-\frac{2}{3}=\frac{15}{21}-\frac{14}{21}=\frac{1}{21}$

14. $\frac{3}{7}-\frac{1}{3}=\frac{9}{21}-\frac{7}{21}=\frac{2}{21}$

16. $\frac{4}{5}-\frac{8}{13}=\frac{52}{65}-\frac{40}{65}=\frac{12}{65}$

18. $5\frac{1}{3}-2\frac{2}{3}=\frac{16}{3}-\frac{8}{3}=\frac{8}{3}=2\frac{2}{3}$

20. $4\frac{4}{5}-3\frac{1}{5}=1\frac{3}{5}$

22. $12\frac{5}{8}-6\frac{4}{9}=\frac{101}{8}-\frac{58}{9}=\frac{909}{72}-\frac{464}{72}=\frac{445}{72}=6\frac{13}{72}$

24. $3\frac{1}{3}-1\frac{7}{8}=\frac{10}{3}-\frac{15}{8}=\frac{80}{24}-\frac{45}{24}=\frac{35}{24}=1\frac{11}{24}$

26.
$$\begin{array}{rcr} 19\frac{5}{6} & = & 19\frac{15}{18} \\ -\underline{12\frac{5}{9}} & = & -\underline{12\frac{10}{18}} \\ & & 7\frac{5}{18} \end{array}$$

28. $1-\frac{1}{12}=?$

$\frac{12}{12}-\frac{1}{12}=?$

$\frac{11}{12}=?$

She used $\frac{11}{12}$ of the wooden stud throughout the day.

30. $56\frac{8}{13}-12\frac{1}{6}=?$

$44\frac{48-13}{78}=?$

$44\frac{35}{78}=?$

The tailor has $44\frac{35}{78}$ meters of fabric left.

32. potassium $\Rightarrow \frac{1}{2}+?=1$

$\frac{1}{2}+?=\frac{2}{2}$

$?=\frac{1}{2}$

sulphur $\Rightarrow \frac{1}{3}+?=1$

$\frac{1}{3}+?=\frac{3}{3}$

$?=\frac{2}{3}$

carbon $\Rightarrow \frac{2}{8}+?=1$

$\frac{2}{8}+?=\frac{8}{8}$

$?=\frac{6}{8}=\frac{3}{4}$

To make 3 kg of gunpowder you would require $\frac{1}{2}$ kg of potassium nitrate, $\frac{2}{3}$ kg of sulphur and $\frac{3}{4}$ kg of carbon.

Summary Test Chapter 2

2. 4 numerator, 7 denominator

4. Proper fraction

6. Improper fraction

8. $\frac{33}{8}=33 \div 8=4\frac{1}{8}$

10. $2\frac{1}{9}=\frac{19}{9}$ $(2 \times 9)+1=19$

12. $4\frac{1}{5}=\frac{21}{5}$ $(4 \times 5)+1=21$

14. $\frac{5}{8}=\frac{2}{3}$ $5 \times 3=15$
$8 \times 2=16$

$15 \neq 16$
False

16. 1, 3, 5, 15.

18. 1, 11.

20. $2=\mathbf{2} \times 1$ $4=\mathbf{2} \times 2$ $6=\mathbf{2} \times 3$
$8=\mathbf{2} \times 4$ $10=\mathbf{2} \times 5$ $12=\mathbf{2} \times 6$
$14=\mathbf{2} \times 7$

22. The first six prime numbers are 2, 3, 5, 7, 11, and 13.

24. $\frac{26}{2}=\mathbf{13}$

$26=2 \times 13$

26. $\frac{81}{3}=27$ $\frac{27}{3}=9$ $\frac{9}{3}=\mathbf{3}$

$81=3 \times 3 \times 3 \times 3$

28. $8=2 \times 2 \times 2$
$9=3 \times 3$

$\text{LCM}=2 \times 2 \times 2 \times 3 \times 3=72$

30. $5\frac{4}{5} \times 1\frac{3}{5}=\frac{29}{5} \times \frac{8}{5}=\frac{232}{25}=9\frac{7}{25}$

32. $\frac{2}{5} \times \frac{5}{2}=\frac{10}{10}=1$

The reciprocal of $\frac{2}{5}$ is $\frac{5}{2}$.

34. $2\frac{1}{2} \div 2\frac{5}{8}=\frac{5}{2} \div \frac{21}{8}=\frac{5}{2} \times \frac{8}{21}=\frac{40}{42}=\frac{20}{21}$

36. $2\frac{1}{2}+3\frac{5}{8}=5\frac{4+5}{8}=5\frac{9}{8}=6\frac{1}{8}$

38.
$$\begin{array}{rcrr} 25\frac{1}{3} & = & 24\frac{4}{3}= & 24\frac{12}{9} \\ \underline{-5\frac{7}{9}} & = & \underline{-5\frac{7}{9}}= & \underline{-5\frac{7}{9}} \\ & & & 19\frac{5}{9} \end{array}$$

40. $5 \div 9=\frac{5}{9}$

Each piece of rope would be $\frac{5}{9}$ meters long.

42. $500 \div 1\frac{1}{3}=500 \div \frac{4}{3}$
$=500 \times \frac{3}{4}$
$=\frac{1500}{4}=375$

The truck can safely hold 375 blocks.

# CHAPTER 3: Decimal Numbers

## Exercise 3.2

2. tenths

(first digit to the right of the decimal point)

4. ten millionths

(seventh digit to the right of the decimal point)

## Exercise 3.4

2. Not a power of 10 because it has more that one '1'.

4. $10^8$ because there is only one '1' and there are 8 zeroes after the '1', and the number is greater than 1.

6. $10^{-4}$ because there is only one '1' and there are 3 zeroes after the decimal point but before the '1'.

8. $10^1$ because there is only one '1' and there is 1 zero after the '1', and the number is greater than 1.

10. $10^2 = 10 \times 10$

$= 100$

## Exercise 3.5

2. $41.6243 = 41.6|243$

Because the digit following the cut-off point is $< 5$, then $41.6243 = 41.6$ (to 1 decimal place).

4. $4999.959 = 4999.9|59$

Because the digit following the cut-off is  5, increase the digit before the cut-off by 1.

$4999.959 = 5\,000.0$
(to one decimal point)

6. $189.76 = 189|.76$

Because the digit following the cut-off point is  5, then increase the digit before the cut-off by 1.

$189.76 \doteq 190$
(to the nearest whole number)

8. $9924.300\,002 = 9924.300\,|\,002$

Because the digit following the cut-off is $< 5$, then just remove the digits following the cut-off.

$9924.300\,002 = 9924.300$
(to 3 decimal places)

10. Cost of gasoline = $68.7 \times 48.9$¢

=3359.43¢

Linda paid 3359¢ for the gasoline, or $33.59.

## Exercise 3.6.1

2.

$$\begin{array}{r} 1.1 \\ +\ \underline{0.2} \\ 1.3 \end{array}$$

4.

$$\begin{array}{r} 3.00 \\ +\ \underline{2.25} \\ 5.25 \end{array}$$

6.

$$\begin{array}{rr} & 0.0003 \\ + & \underline{22.1000} \\ & 22.1003 \end{array}$$

8.

$$\begin{array}{rr} & 12.1000 \\ + & 13.4440 \\ + & 100.0001 \\ + & \underline{985.3000} \\ & 1110.8441 \end{array}$$

Exercise 3.6.2

2.

$$\begin{array}{rr} & 49.68 \\ - & \underline{19.79} \\ & 29.89 \end{array}$$

4.

$$\begin{array}{rr} & 2.0000 \\ - & \underline{0.0001} \\ & 1.9999 \end{array}$$

6.

$$\begin{array}{rr} & 13.5 \\ - & \underline{8.0} \\ & 5.5 \end{array}$$

7. 0.0034

Exercise 3.6.3

2. $209 \times 1000 = 209.000 \times 1000$
$= 209000$

Move decimal point 3 places to the right.

4. $0.003 \times 100000$
$= 0.00300 \times 100000$
$= 300$

Move the decimal point 5 places to the right.

6. $822.003 \times 0.000000001$

$= 00000000822.003 \times 0.000000001$
$= 0.000000822003$

Move the decimal point 9 places to the left.

Exercise 3.6.4

2.

$$\begin{array}{rr} & 165 \\ \times & \underline{12} \\ & 330 \\ & \underline{1650} \\ & 1980 \end{array}$$

Since there is a total of 1 digit after the decimal point in the question,
$16.5 \times 12 = 198.0$

4.

$$\begin{array}{rr} & 15 \\ \times & \underline{4} \\ & 60 \end{array}$$

Since there is a total of 7 digits after the decimal point in the question,
$1.5 \times 0.000\,004 = 0.000\,006$

6.

$$\begin{array}{rr} & 3937 \\ \times & \underline{65} \\ & 19685 \\ & \underline{236220} \\ & 255905 \end{array}$$

A rope that is 6.5 metres long is also 255.905 inches

8.

$$\begin{array}{rr} & 500 \\ \times & \underline{68453} \\ & 1500 \\ & 25000 \\ & 200000 \\ & 4000000 \\ & \underline{30000000} \\ & 34226500 \end{array}$$

There are 5 digits in total after the decimal point in the question, so there will be 5 digits after the decimal point in the answer: $342.265000
That is, Stacey will receive US$342.27 (rounding to the nearest cent)

Exercise 3.6.5

2. Move the decimal point 2 places to the left because $100 = 10^2$: 7.32664.

4. Move the decimal point 3 places to the right because $0.001=10^{-3}$: 2002.

6. Move the decimal point 6 places to the right because $0.000\ 001 = 10^{-6}$. Note that you need to add extra zeroes to the end of 85.6 so that the decimal point can be moved: 85 600 000.

8. Move the decimal point 4 places to the left because $10\ 000 = 10^4$. Note that you need to add extra zeroes in front of 2.6 so that the decimal point can be moved: 0.00 026.

Exercise 3.6.6

2.

$$4\overline{)0.22500}\quad \text{quotient } 0.05625$$

4.

$$8\overline{)2.4000}\quad \text{quotient } 0.3$$

6. $5 \div 0.0005 = 50\ 000 \div 5$

$$5\overline{)50000.0}\quad \text{quotient } 10000$$

8.

$489\ 127.1 \div 0.0\ 0\ 7$
$= 4\ 891\ 271\ 000 \div 7$

$$7\overline{)4891271000}\quad \text{quotient } 698753000$$

10.

235 kilometres $\div$ 2.5 hours
= 2350 kilometres $\div$ 25 hours

$$25\overline{)2350}\quad \text{quotient } 94$$

Joseph travelled at an average speed of 94 kilometres per hour.

Exercise 3.6

2.

$$\begin{array}{rr} & 246 \\ \times & \underline{127} \\ & 1722 \\ & 4920 \\ & \underline{24600} \\ & 31242 \end{array}$$

Because there are 2 digits after the decimal point in the multiplication, there are 2 after the decimal point in the answer. $24.6 \times 12.7 = 312.42$.

4.

$$\begin{array}{rr} & 8.000 \\ - & \underline{4.002} \\ & 3.998 \end{array}$$

6.

$$\begin{array}{rr} & 72.600 \\ + & 1000.000 \\ + & \underline{0.004} \\ & 1072.604 \end{array}$$

8. Move the decimal point 4 places to the left because $0.0001 = 10^{-4}$. Note that you need to add extra zeroes in front of 42.3 so that the decimal point can be moved: 0.00423.

$$15\overline{)146.7480} \quad 9.7832$$

Exercise 3.7

2.

$$3\overline{)2.000000000000} \quad 0.66666666...$$

$$\tfrac{2}{3} = 0.\dot{6}$$

4.

$$100\overline{)57.000} \quad 0.57$$

$$\tfrac{57}{100} = 0.57$$

6.

$$15\overline{)8.000000000000} \quad 0.53333333...$$

$$\tfrac{8}{15} = 0.5\dot{3}$$

8.

$$4\overline{)1.00} \quad 0.25$$

$$\tfrac{1}{4} = 0.25$$

10. $\frac{867}{1000} = 867.0 \div 1000$

$= 0.867$

Move the decimal point 3 places to the left because $1000 = 10^3$.

Exercise 3.8

2. There is 1 digit in the repeat, but there is 1 extra digit before the repeat starts. Since 1 + 1 =2, the power of ten is 2; we will have to multiply the decimal number by $10^2 = 100$. We will also have to multiply the original number by $10^1 = 10$ because of the extra digit before the repeat starts.

$12.1666666... \times 100 = 1216.6666666...$
$12.1666666... \times 10 = 121.66666666...$
$1216.66666... - 121.66666... = 1095$

Subtract the powers of 10:
$100 - 10 = 90$

Divide 1095 by 90:

$$\frac{1095}{90}$$

$$= \frac{365}{30}$$

$$= \frac{73}{6}$$

$$= 12\tfrac{1}{6}$$

4.

There are 6 digits in the repeat, so the power of ten is 6; we will have to multiply the decimal number by $10^6 = 1\ 000\ 000$.

$0.428571428571428571... \times 1000000$
$= 428\ 571.428\ 571\ 428\ 571...$

$$\begin{array}{r} 428571.428571428571... \\ -\ \underline{0.428571428571428571..} \\ 428571 \end{array}$$

Subtract 1 from 1 000 000:
1 000 000 - 1 = 999 999

Divide 428 571 by 999 999:

$\frac{428571}{999999} = \frac{142857}{333333}$ (Dividing by 3)

$= \frac{47619}{111111}$ (Dividing by 3)

$= \frac{15873}{37037}$ (Dividing by 3)

$= \frac{1443}{3367}$ (Dividing by 11)

$= \frac{111}{259}$ (Dividing by 13)

$= \frac{3}{7}$ (Dividing by 37)

$0.428571\ 428571\ 428571... = \frac{3}{7}$

6. There is 1 digit in the repeat, so the power of ten is 1; we will have to multiply the decimal number by $10^1=10$.

$0.88888... \times 10 = 8.888888...$
$8.88888... - 0.88888... = 8$

$10 - 1 = 9$

Divide 8 by 9: $\frac{8}{9}$
$0.88888... = \frac{8}{9}$

Exercise 3.9

2. Move the decimal point 5 places to the left: 0.000027

4. Move the decimal point 2 places to the right: 210

6. The decimal point has to be moved 15 places to the left to express the number in scientific notation: $7.95 \times 10^{15}$

8. The decimal point has to moved 3 places to the right to express the number in scientific notation: $2.37 \times 10^3$

10. The decimal point has to be moved 1 place to the left to express the number in scientific notation: $2.24 \times 10^1$

12. The decimal point has to moved 1 place to the left to express the number in correct scientific notation, so add 1 to the power of 10: $4.36 \times 10^5$

14. The decimal point has to be moved 2 places to the right to express the number in correct scientific notation, so subtract 2 from the power of 10:
$2.3 \times 10^{-4}$

16.

$9.2 \times 10^{-5} \times 7.3 \times 10^{-5}$

$= 9.2 \times 7.3 \times 10^{-5} \times 10^{-5}$
$= 67.16 \times 10^{-5+-5}$
$= 67.16 \times 10^{-10}$
$= 6.716 \times 10^{-9}$

18.

$7.1 \times 10^{-3} \times 1.3 \times 10^{3}$

$= 7.1 \times 1.3 \times 10^{-3} \times 10^{3}$
$= 9.23 \times 10^{-3+3}$
$= 9.23 \times 10^{0}$
$= 9.23 \times 1$
$= 9.23$

20.

$(6.4 \times 10^{-7}) \div (1.6 \times 10^{5})$

$= (6.4 \div 1.6) \times (10^{-7} \div 10^{5})$
$= 4.0 \times 10^{-7-5}$
$= 4.0 \times 10^{-12}$

22.

$(4.006 \times 10^{-12}) \div (1.3 \times 10^{-18})$

$= (4.006 \div 1.3) \times (10^{-12} \div 10^{-18})$
$= 3.0815385 \times (10^{-12--18})$
$= 3.0815385 \times 10^{6}$
$= 3081538.5$

24.

Because the decimal point has to moved 8 places to the left, 100 000 000 ° C
$= 1.0 \times 10^{8}$ ° C

26.

$6.0 \times 10^{-5} \times 1.9 \times 10^{5}$
$6.0 \times 1.9 \times 10^{-5} \times 10^{5}$
$= 11.4 \times 10^{-5+5}$
$= 11.4 \times 10^{0}$
$= 11.4 \times 1$
$= 11.4$

The strip of hair would be 11.4 metres wide.

End of Chapter 3 Test

2.

a.

$$\begin{array}{rr} & 35.67000 \\ + & 2.01560 \\ + & \underline{0.00006} \\ & 37.68566 \end{array}$$

b.

$$\begin{array}{rr} & 8002 \\ \times & \underline{465} \\ & 40010 \\ & 480120 \\ & \underline{3200800} \\ & 3720930 \end{array}$$

Because there is a total of 5 digits after the decimal point in the question, then
$8.002 \times 4.65 = 37.2093$

c. Because $0.001 = 10^{-3}$, then the decimal point needs to moved 3 places to the left:
$12.056 \times 0.001 = 0.012056$

d. $84.7 \div 0.007 = 84\,700 \div 7$

$= 12\,100$

e. $49 \div 1000 = 49 \div 10^{3}$

$=0.049$

Move the decimal point 3 places to the left.

f.

$$\begin{array}{rr} & 9.600 \\ - & \underline{7.968} \\ & 1.632 \end{array}$$

4.a. Place the cut-off indicator in the appropriate place: 45.6|892
Because the digit immediately to the right of the cut-off is greater than 4, increase the digit to the left of the cut-off by 1: 45.7

b. Place the cut-off indicator in the appropriate place: 9.999|8
Because the digit immediately to the right of the cut-off is greater than 4, increase the digit to the left of the cut-off by 1. But, because the digit to the left of the cut-off is 9, this increases to 10; the zero is written down, and the '1' is carried to the next digit: 10.000

c. Place the cut-off indicator in the appropriate place: 0.0048|3
Because the digit immediately to the right of the cut-off is less than 5, just discard the digits to the right of the cut-off: 0.0048

6. The decimal point needs to be moved 11 places to the left so
$256\ 000\ 000\ 000 = 2.56 \times 10^{11}$

b. The decimal point needs to moved 7 places to the right, so
$0.000\ 000\ 34 = 3.4 \times 10^{-7}$

c. The decimal point needs to be moved 2 places to the right, so
$0.03 = 3.0 \times 10^{-2}$

# CHAPTER 4: Ratio and Proportions

## Exercise 4.1

2. $\frac{15}{27}=\frac{5\times3}{9\times3}=\frac{5}{9}$

4. $\frac{1}{2}$

6. $\frac{9}{13}$

8. $\frac{10}{14}=\frac{5}{7}$

10. $\frac{8}{32}=\frac{1}{4}$

## Exercise 4.2

2. $$\frac{6\ \text{l}}{100\ \text{km}}=\frac{60\ \text{l}}{x\ km}$$

$6\ \text{l}\ \times x\ \text{km} = 60\ \text{l}\ \times\ 100\text{km}$

$x\ \text{km} = \frac{60\ \text{l}\times 100\text{km}}{6\ \text{l}}$

$x\ \text{km} = 10\ \times\ 100\ \text{km}$

$x\ \text{km} = 1000\ \text{km}$

Joe's car will travel 1000 kilometres per tankful.

4. $$\frac{100\ \text{ml}}{1\ \text{m}}=\frac{1000\ \text{ml}}{x\ \text{m}}$$

$100\ \times x = 1000\ \times\ 1$

$x = \frac{1000}{100}$

$x\ =\ 10$

10 meters of wood can be glued together with 1000 ml of glue.

## Exercise 4.3

2. $\frac{2}{x}=\frac{16}{32}$

$2\times32 = 16\times x$

$64 = 16x$

$\frac{64}{16}=x$

$4=x$

4. $\frac{5}{28}=\frac{y}{56}$

$5\times56 = 28y$

$280 = 28y$

$10=y$

6. $\frac{v}{55}=\frac{4}{110}$

$4\times55 = 110v$

$220 = 110v$

$2=v$

8. $\frac{4}{3}=\frac{20}{c}$

$4c = 20\times3$

$4c = 60$

$c=\frac{60}{4}$

$c=15$

10. $\frac{x}{0.2} = \frac{6}{6.4}$

$6.4x = 6 \times 0.2$

$6.4x = 1.2$

$x = \frac{1.2}{6.4}$

$x = 0.1875$

12. $\frac{3}{8} = \frac{t}{24.8}$

$3 \times 24.8 = 8t$

$74.4 = 8t$

$\frac{74.4}{8} = t$

$9.3 = t$

Exercise 4.4

2. $\frac{4\text{ l}}{1\text{ tree}} = \frac{200\text{ l}}{x\text{ tree}}$

$4x = 200$

$x = \frac{200}{4}$

$x = 50$

50 trees are need to produce 200 litres per day.

4. $\frac{17\text{ welds}}{1\text{ m}} = \frac{x\text{ welds}}{7\text{ m}}$

$x = 17 \times 7$

$x = 119$

119 spot welds are on the truck

6. $\frac{1}{24} = \frac{x}{6}$

$6 \times 1 = x \times 24$

$6 = 24x$

$\frac{6}{24} = x$

$\frac{1}{4} = x$

The length of the scale model is $\frac{1}{4}$ of a metre.

Summary Test Chapter 4

2. 1 to 9

4. 1 to 5

6. $\frac{3}{5}$

8. $\frac{12}{240} = \frac{1}{20}$

10. $\frac{12}{2} = \frac{x}{10}$

$12 \times 10 = 2x$

$120 = 2x$

$x = 60$

12. $\frac{10}{56} = \frac{x}{112}$

$56x = 112 \times 10$

$56x = 1120$

$x = \frac{1120}{56}$

$x = 20$

14.

The ratio of revolutions per second is 400 to 1.

16. $\frac{1000 \text{ m}}{x \text{ m}} = \frac{1 \text{ m}}{40 \text{ m}}$

$x = 40 \times 1000$

$x = 40000$

The pilot will *NOT* be able to land at the airport since she can only glide 40 kilometres, and the airport is 46 kilometres away.

18. $\frac{1 \text{ kg}}{120 \text{ cups}} \rightarrow \frac{360 \text{ cents}}{120 \text{ cups}} = \frac{3\text{cents}}{1 \text{ cup}}$

Profit per cup = sell price - cost price

= 90 - 3

= 87 cents

$120\text{cups} \times \frac{87 \text{ cents}}{\text{cup}} = 10440 \text{ cents}$

$= \$104.40$

Mr. Valdez makes $104.40 per kilogram (120 cups) of coffee sold.

20. $\frac{a}{b} = \frac{7}{9}$

$\frac{a}{22.5} = \frac{7}{9}$

$9a = 22.5 \times 7$

$9a = 157.5$

$a = \frac{157.5}{9}$

$a = 17.5$

There are 17.5 a's.

22.

New: $\frac{\$13000}{200000 \text{ km}} = \frac{\$0.065}{1 \text{ km}}$

Used: Kilometres left on used car

= 200000 - 55000=145000 km

$\frac{\$8000}{145000} = \frac{\$0.055}{1 \text{ km}}$

Based solely on the cost per kilometre, the used car should be purchased.

24. $\frac{36 \text{ L}}{60 \text{ min}} = \frac{x \text{ L}}{200 \text{ min}}$

Note: 3 hour and 20 minutes is equal to 200 minutes.

$60x = 36 \times 200$

$x = \frac{36 \times 200}{60}$

$x = 120$

120 litres would be used on a 3 hour and 20 minute flight.

# CHAPTER 5: Percentages

## Exercise 5.1

2. $25.6\% = \frac{25.6}{100} = 0.256$

4. $89.54\% = \frac{89.54}{100} = 0.8954$

6. $1\frac{1}{2}\% = \frac{3}{2}\% = \frac{\frac{3}{2}}{100} = \frac{3}{2} \times \frac{1}{100}$

$= \frac{3}{200} = 0.015$

8. $11.5\% = \frac{11.5}{100} = 0.115$

10. $129\% = \frac{129}{100} = 1.29$

12. $0.05\% = \frac{0.05}{100} = 0.0005$

14. $12\frac{2}{3}\% = \frac{38}{3}\% = \frac{\frac{38}{3}}{100} = \frac{38}{3} \times \frac{1}{100}$

$= \frac{38}{300} = 0.12\dot{6}$

## Exercise 5.2

2. $2.5 = \frac{250}{100} = 250\%$

4. $0.65 = \frac{65}{100} = 65\%$

6. $0.125 = \frac{12.5}{100} = 12.5\%$

8. $0.13 = \frac{13}{100} = 13\%$

10. $0.898 = \frac{89.8}{100} = 89.8\%$

12. $0.029 = \frac{2.9}{100} = 2.9\%$

## Exercise 5.3

2. $\frac{3}{5} = \frac{60}{100} = 60\%$

4. $\frac{3}{4} = \frac{75}{100} = 75\%$

6. $\frac{7}{10} = \frac{70}{100} = 70\%$

8. $\frac{3}{2} = \frac{150}{100} = 150\%$

10. $\frac{2}{13} \doteq \frac{0.1538}{100} \doteq 15.38\%$

12. $\frac{5}{6} = \frac{0.8\dot{3}}{100} = 83.\dot{3}\%$

## Exercise 5.4

2. $25\% = \frac{25}{100} = \frac{1}{4}$

4. $13.5\% = \frac{13.5}{100}$

6. $\frac{1}{3}\% = \frac{\frac{1}{3}}{100} = \frac{1}{3} \times \frac{1}{100} = \frac{1}{300}$

8. $66\frac{2}{3}\% = \frac{\frac{200}{3}}{100} = \frac{200}{3} \times \frac{1}{100} = \frac{200}{300} = \frac{2}{3}$

10. $489\% = \frac{489}{100}$

12. $33\frac{1}{3}\% = \frac{\frac{100}{3}}{100} = \frac{100}{3} \times \frac{1}{100} = \frac{1}{3}$

Exercise 5.5

2. a) 100% - 11.6% = 88.4%

88.4% left after the decrease

$\frac{88.4}{100} = \frac{x}{14250}$

x = $12597

The value of the fund after the decrease is $12597.

b) $\frac{156.5}{100} = \frac{x}{12597}$

$x$ = $19714.31

The value of the fund after the increase is $19714.31.

4. $\frac{125}{100} = \frac{x}{500}$

$x = 625$

625 - 500 = 125

You now earn $125 more.

6. $\frac{128}{100} = \frac{x}{500}$

x = 640

You will receive an additional 140 grams of cereal.

8. a) 100% - 12.5% = 87.5%

$\frac{87.5}{100} = \frac{x}{12250}$

x = 10718.75

The value of the fund after the decrease is $10718.75.

b) $\frac{126.5}{100} = \frac{x}{10718.75}$

$x \doteq 13559.22$

The value of the fund after the increase is now $13559.22 .

Exercise 5.6

2. $\frac{115}{100} = \frac{x}{45.56}$

$x \doteq \$52.39$

The total amount of the bill, including tip, would be $52.39.

4. 100% - 35% = 65%

$\frac{65}{100} = \frac{x}{75}$

$x = \$48.75$

The shoes would cost $48.75.

6. $\frac{115}{100} = \frac{26.45}{x}$

$x = \$23$

The bill prior to the tip was $23.00.

8. $\$23063 \times \frac{3.6}{100} \doteq \$830.27$

Your addition cost of living will increase by $830.27.

10. $\frac{50}{100} = \frac{x}{2520}$

$x = 1260$

The computer will be worth $1260 at the end of the year.

Summary Test Chapter 5

2. $33.5\% = \frac{33.5}{100} = 0.335$

4. $0.85 = \frac{85}{100} = 85\%$

6. $1.54 = \frac{154}{100} = 154\%$

8. $2\frac{1}{2} = \frac{5}{2} = \frac{250}{100} = 250\%$

10. $68\frac{2}{3}\% = \frac{\frac{206}{3}}{100} = \frac{206}{3} \times \frac{1}{100} = \frac{206}{300}$

12. $1140\% = \frac{1140}{100}$

14. $\frac{107}{100} = \frac{350}{x}$

$x = 327.1$

350 - 327.1 = 22.9

The increase was $22.90.

# CHAPTER 6: Metric System

Exercise 6.3

2. Deci means $\frac{1}{10}$, so there are 10 dm in 1 metre.

4. Kilo means 1000, so 1 Kilogram = 1000grams = 10 000 grams

Therefore 1 000 000 grams is larger than 10 kilograms.

6. First, convert kilometres to metres: 25 kilometres = 25 000 meters.

Then, convert metres to millimetres. There are 1000 millimetres in 1 metre, so:

$$25\,000 \text{ metres} = 25\,000 \times 1000 \text{ mm}$$
$$= 25\,000\,000 \text{ mm}$$

8. Kilo means 1 000, so 87 kilo joules = 87 000 joules.

10. Milli means $\frac{1}{1000}$, so there are 1000 millivolts in 1 volt.

12. 4 kg = 4000 g

$4000 \div 250$

$= \frac{4000}{250}$

$= \frac{16}{1}$

So, I am making 16 times the amount of the recipe. I will need to microwave the beans for 2 minutes $\times$ 16 = 32 minutes.

14. 2m = 200 cm

$200 \text{ cm} \div 5 = 40 \text{ cm}$

Metres are converted to centimetres because the "amount" will be a larger number, which can then be divided into pieces.

16. There are 10 millilitres in 1 centilitre, so 75 cL = 75 $\times$ 10 mL = 750 mL.

18. 'K' is used as an abbreviation for 'kilo', and kilo means 1000.
This means that the salary is \$32 $\times$ 1000 = \$32 000.

20. In 4 hours of flying, the plane uses

$13\,500\text{L} \times 4 = 54\,000\text{L}$
$1 \text{ L} = 0.000\,001 \text{ ML}$
$54\,000\text{L} = 54\,000 \times 0.000\,001$
$= 0.054 \text{ ML}$

The plane uses 0.054 megalitres in 4 hours.

Exercise 6.4

2. 64 kilobytes $\approx$ 64 000 bytes.
16 megabytes $\approx$ 16 000 000 bytes.

16 megabytes $\div$ 64 kilobytes
$\approx$ 16 000 000 $\div$ 64 000 = 250

The modern computer's memory is 250 times larger than those of the early computes.

4. 3.5 kg = 3 500 grams
3 500 grams $\div$ 180 grams =
$\frac{3\,500}{180}$

$\frac{3\,500}{180} = \frac{350}{18} = 19\frac{4}{9}$

Jerri can invite 19 people to dinner.

6. 45 L = 45 000 millilitres
$45\,000 \div 600 = \frac{45\,000}{600}$

$= \frac{450}{6}$

$= 75$

I will need 75 bottles.

Exercise 6.5

2. British 1 pint = 20 fluid ounces.

$$\begin{aligned} 5 \text{ pints} &= 5 \times 20 \text{ fluid ounces} \\ &= 100 \text{ fluid ounces} \end{aligned}$$

1 dose = 4 fluid ounces, so the number of doses is $100 \div 4 = 25$.

American: 1 pint = 16 fluid ounces

$$\begin{aligned} 5 \text{ pints} &= 5 \times 16 \text{ fluid ounces} \\ &= 80 \text{ fluid ounces} \end{aligned}$$

1 dose = 4 fluid ounces, so the number doses is $80 \div 4 = 20$.

If the amount of Snake Oil was measure in British pints, there are 25 does; if it is American, there are 20 doses.

4. British: 1 ton = 2240 pounds

$$\begin{aligned} 5 \text{ tons} &= 5 \times 2240 \text{ pounds} \\ &= 11200 \text{ pounds} \end{aligned}$$

1 block = 5 pounds, so the number of blocks = $11\,200 \div 5 = 2\,240$.

American: 1 ton = 2 000 pounds

$$\begin{aligned} 5 \text{ tones} &= 5 \times 2\,000 \text{ pounds} \\ &= 10\,000 \text{ pounds} \end{aligned}$$

1 block = 5 pounds, so the number of blocks = $10\,000 \div 5 = 2\,000$.

Mohammed's truck can carry 2 240 blocks if measured in British tons, or 2 000 blocks if American tons.

Exercise 6.6

2.a. 1 mile = 1.61 km, so

$$\begin{aligned} 35 \text{ miles} &= 35 \times 1.61 \text{ kilometres} \\ &= 56.35 \text{ kilometres} \end{aligned}$$

35 miles per hour is approximately 56 km per hour (or 60 kilometres per hour).

b. 1 mile = 1.61 km, so

$$\begin{aligned} 20 \text{ miles} &= 20 \times 16.1 \text{ kilometres} \\ &= 32.2 \text{ kilometres} \end{aligned}$$

20 miles per hour is approximately 32 km per hour (or 30 kilometres per hour).

c. 1 mile = 1.61 km, so

$$\begin{aligned} 55 \text{ miles} &= 55 \times 1.61 \text{ kilometres} \\ &= 88.55 \text{ kilometres} \end{aligned}$$

35 miles per hour is approximately 89 km per hour (or 90 kilometres per hour).

Exercise 6.7

2.a. Exact

$$80° - 32° = 48°$$

$$48° \div 9 = 5.33°$$

$$5.33° \times 5 = 26.7° \text{ C}$$

$$80° \text{ F} = 26.7° \text{ C}$$

Approximate

$$80° - 30° = 50°$$

$$50° \div 2 = 25° \text{ C}$$

$$80° \text{ F} = 25° \text{ C}$$

b. Exact

$$300° - 32° = 268°$$

$$268° \div 9 = 29.78°$$

$$29.78° \times 5 = 148.9° \text{ C}$$

$$300° \text{ F} = 148.9° \text{ C}$$

Approximate

$300° - 30° = 270°$

$270° \div 2 = 135° C$

$300° F = 135° C$

c. Exact

$600° - 32° = 568°$

$568° \div 9 = 63.11°$

$63.11° \times 5 = 315.6° C$

$600° F = 315.6° C$

Approximate

$600° - 30° = 570°$

$570° \div 2 = 285° C$

$600° F = 285° C$

d. Exact

$60° - 32° = 28°$

$28° \div 9 = 3.11°$

$3.11° \times 5 = 15.6° C$

$60° F = 15.6° C$

Approximate

$60° - 30° = 30°$

$30° \div 2 = 15° C$

$60° F = 15° C$

4. Put on your coat; temperature is probably in Fahrenheit because the plane is from the USA, and 35° is just above freezing (32° F is the freezing point of water).

Exercise 6.8

2. 1 mL of water has a mass of 1 g
1000 mL of water has a mass of 1000g
that is, 1 L of water has a mass of 1 Kg
90 L of water has mass of 90 kg.
Therefore, a full 90 L tank of gas has an additional 90 kg of mass.

4. 1 mL of water has a mass of 1 g
1000 mL of water has a mass of 1000g
that is, 1 L of water has a mass of 1 kg
In 8 hours, the plane uses 108 000 L of fuel.

108 000 L of water has a mass of
108 000 kg (or 108 tonnes).
Therefore, the plane requires 108 000 kg of fuel to be loaded for an 8 hour flight.

End of Chapter 6 Test

2. Total mass of people: 50 kg + 90 kg + 120 kg + 76 kg + 20 kg + 93 kg + 87 kg + 15 kg = 551 kg

Total mass of people with the refrigerator: 551 kg + 650 kg = 1201 kg

1.4 tonnes = 1.4 × 1000 kg
= 1400 kg

The capacity of the elevator was not exceeded.

4. 1 Calorie = 4 185 joules

2000 Calories = 2000 × 4 185 joules
=8 370 000 joules
1000 joules = 1 kilojoule, so 8 370 000 joules = 8 370 000 ÷ 1000 joules
= 8 370 kilo joules

6. British: 1 pint = 20 fluid ounces

1 gallon = 8 pints
= 8 × 20 fluid ounces
= 160 fluid ounces

American: 1 pint = 16 fluid ounces

1 gallon = 8 pints
= 8 × 16 fluid ounces
= 128 fluid ounces

8. 1 kilogram = 2.2 pounds

59 cents per pound
= 2.2 × 59 cents per kilogram
= 129.8 cents per kilogram

The bananas should have been 130 cents per kilogram, or $1.30 per kg.

The checkout price was incorrect.

10.a.

32 ÷ 5 = 6.4
6.4 × 9 = 57.6
57.6 + 32 = 89.6

32° C = 89.6° F

b.

100 ÷ 5 = 20
20 × 9 = 180
180+32 = 212

100° C = 212° F

c.

500 ÷ 5 = 100
100 × 9 = 900
900 + 32 = 932

500° C = 932 ° F

12. Volume of mulch required:

5 cm × 20 m × 5 m

Convert all measures to the same unit:

= 5 cm × 2000 cm × 500 cm
= 5 000 000 $cm^3$
1 $cm^3$ = 1 mL
1000$cm^3$ = 1 L
5 000 000 $cm^3$ = 5 000 L

The mulch is sold in 50L bags, so I will need to buy: 5 000 ÷ 50 = 100 bags.

# CHAPTER 7: Algebra and Formulae

Exercise 7.2.1

2. Variable: $s$ (because it is unknown)
Coefficient: 5 (because it multiplies a variable)
Constant: 4 (because it is a number without a variable)

4. Variable: $y$ (because it is unknown)
Coefficient: $-3$ (because it multiplies a variable)
Constant: 18 (because it is a number without a variable)

Exercise 7.2.2

2. $5y - 8y = (5-8)y$
$= -3y$

4. $8K + 5T - 12T - 20K$
$= 8K - 20K + 5T - 12T$
$= (8-20)K + (5-12)T$
$= -12K - 7T$

6. $2(x+y) - 5(x-y)$
$= 2x + 2y - 5x + 5y$
$= 2x - 5x + 2y + 5y$
$= -3x + 7y$

8. $8x \div 2$
$= \frac{8x}{2}$
$= 4x$

10. $10K \div 10$
$= \frac{10K}{10}$
$= K$

Exercise 7.3

2. $I = \frac{PRT}{100}$

$= \frac{\$5000 \times 5 \times 10}{100}$

$= \frac{\$250000}{100}$

$= \$2500$

The simple interest on \$5000, invested at 5% for 10 years is \$2500.

4. $M = mv$
$= 8 \times 10$ kg metres per second
$= 80$ kg metres per second

The momentum of the body is 80 kg $\frac{\text{m}}{\text{sec}}$.

6. $E = mc^2$

$= 0.0025 \times (300000)^2$
$= 0.0025 \times 90000000000$
$= 225000000 \text{ kg } \frac{\text{m}^2}{\text{s}^2}$
$= 2.25 \times 10^8$ joules

The energy of the ping pong ball's atoms would be equal to
$2.25 \times 10^8 \text{ kg}\frac{\text{m}^2}{\text{s}^2}$.

8. a. $C = \frac{5}{9}(F - 32)$
$= \frac{5}{9}(60 - 32)$
$= \frac{5}{9} \times 28$
$= \frac{140}{9}$
$\approx 15.56$

$60°F \approx 15.56°C$

b. $C = \frac{5}{9}(F - 32)$
$= \frac{5}{9}(350 - 32)$
$= \frac{5}{9} \times 318$
$= \frac{1590}{9}$
$\approx 176.67$

$350°F \approx 176.67°C$

c. $C = \frac{5}{9}(F - 32)$

$= \frac{5}{9}(-40 - 32)$

$= \frac{5}{9} \times (-72)$

$= \frac{-360}{9}$

$= -40$

$-40°F = -40°C$

10. a.
The bacterium splits every 20 minutes, so there was only 1 split.

Number of bacteria $= 2^1$
$= 2$

There are 2 bacteria after 20 minutes.

b.
After 3 hours, there had been 9 splits.

Number of bacteria $= 2^n$
$= 2^9$
$= 512$

After 3 hours, there were 512 bacteria.

c.
After 5 hours, there had been 15 splits.

Number of bacteria $= 2^n$
$= 2^{15}$
$= 32768$

After 5 hours, there were 32768 bacteria.

Exercise 7.4

2. $3t - 5 = 16$

$3t - 5 + 5 = 16 + 5$

$3t = 21$

$\frac{3t}{3} = \frac{21}{3}$

$t = 7$

4. $4k + 4.5 = 8.7$

$4k + 4.5 - 4.5 = 8.7 - 4.5$

$4k = 4.2$

$\frac{4k}{4} = \frac{4.2}{4}$

$k = 1.05$

6. $5 = \frac{y}{9}$

$5 \times 9 = \frac{y}{9} \times 9$

$45 = y$

$y = 45$

8. $6(t - 5) = 7(2t + 1)$

$6t - 30 = 14t + 7$

$6t - 30 + 30 = 14t + 7 + 30$

$6t = 14t + 37$

$6t - 14t = 14t - 14t + 37$

$-8t = 37$

$\frac{-8t}{-8} = \frac{37}{-8}$

$t = \frac{-37}{8}$

10. $\frac{x}{2} + 5 = 7$

$\frac{x}{2} + 5 - 5 = 7 - 5$

$\frac{x}{2} = 2$

$\frac{x}{2} \times 2 = 2 \times 2$

$x = 4$

Exercise 7.5

2. Let the number be $x$

$3x + 10 = 46$

$3x + 10 - 10 = 46 - 10$

$3x = 36$

$\frac{3x}{3} = \frac{36}{3}$

$x = 12$

The number is 12.

4. Let the number of kilometres travelled be $x$.

$$\begin{aligned}90 &= 35.95 + 0.1 \times x\\ 90 - 35.95 &= 35.95 - 35.95 + 0.1x\\ 54.05 &= 0.1x\\ 54.05 \times 10 &= 0.1x \times 10\\ 540.5 &= x\end{aligned}$$

I travelled 540.5 km.

6. Let the length of the shorter piece of rope be $x$.
The length of the longer piece is $4x$.
The total length of the rope is $x + 4x$; this is, $5x$.
But the total length of the rope is 2m.

Therefore, $5x = 2\text{m}$
$5x = 200$ cm
$x = 40$ cm

$4x = 40 \text{ cm} \times 4$
$= 160$ cm

The shorter piece of rope is 40 cm, and the longer piece is 160 cm.

Exercise 7.6

2. Let the number be $x$

$$\begin{aligned}3x &= 4x - 5\\ 3x - 4x &= 4x - 4x - 5\\ -1x &= -5\\ -1x \times (-1) &= -5 \times (-1)\\ x &= 5\end{aligned}$$

The number is 5.

4. Note that twins are the same age.
Let the age of Louisa be L.
The age of Megan = L

$4L + 3L = 49$
$7L = 49$
$\frac{7L}{7} = \frac{49}{7}$
$L = 7$

Louisa and Megan are 7 years old.

6. Let the cost of the jacket be $J$.
Cost of the pants $= \frac{2}{3}J$
Cost of jacket + cost of pants=\$250

$J + \frac{2}{3}J = 25$

$\frac{3}{3}J + \frac{2}{3}J = 250$

$\frac{2+3}{3}J = 250$

$\frac{5}{3}J = 250$

$3 \times \frac{5}{3}J = 250 \times 3$

$5J = 750$

$\frac{5J}{5} = \frac{750}{5}$

$J = 150$

Cost of pants $= \frac{2}{3}J$

$$\begin{aligned}&= \tfrac{2}{3} \times 150\\ &= \tfrac{2\times150}{3}\\ &= \tfrac{300}{3}\\ &= 100\end{aligned}$$

The pants cost \$100, and the jacket cost \$150.

8. Let the regular price of the box of soap be $S$.

$$\begin{aligned}S + \tfrac{1}{2}S &= 4.26\\ \tfrac{2}{2}S + \tfrac{1}{2}S &= 4.26\\ \tfrac{3}{2}S &= 4.26\\ 2 \times \tfrac{3}{2}S &= 2 \times 4.26\\ 3S &= 8.52\\ \tfrac{3S}{3} &= \tfrac{8.52}{3}\\ S &= 2.84\end{aligned}$$

The regular price of the box os soap is \$2.84.

10. Let the original price of the stereo be $P$.

If the price was reduced by 18%, then it sold for 100% - 18% = 82% of the original price.

$$82\% \times P = \$984$$
$$\tfrac{82}{100} \times P = 984$$
$$100 \times \tfrac{82}{100} \times P = 984 \times 100$$
$$82P = 98\,400$$
$$\tfrac{82P}{82} = \tfrac{98400}{82}$$
$$P = 1\,200$$

The original price of the stereo was \$1200.

Exercise 7.7

2. $V = \tfrac{4}{3}\pi r^2 h$
$$3 \times V = 3 \times \tfrac{4}{3}\pi r^2 h$$
$$3V = 4\pi r^2 h$$
$$\tfrac{3V}{4\pi r^2} = \tfrac{4\pi r^2 h}{4\pi r^2}$$
$$\tfrac{3V}{4\pi r^2} = h$$
$$h = \tfrac{3V}{4\pi r^2}$$

4. $E = IR$
$$\tfrac{E}{I} = \tfrac{IR}{I}$$
$$\tfrac{E}{I} = R$$
$$R = \tfrac{E}{I}$$

6. $C = 2\pi r$
$$\tfrac{C}{2\pi} = \tfrac{2\pi r}{2\pi}$$
$$\tfrac{C}{2\pi} = r$$
$$r = \tfrac{C}{2\pi}$$

8. $I = Prt$
$$\tfrac{1}{rt} = \tfrac{Prt}{rt}$$
$$\tfrac{1}{rt} = P$$
$$P = \tfrac{1}{rt}$$

10. $P = EI$
$$\tfrac{P}{E} = \tfrac{EI}{EI}$$
$$\tfrac{P}{E} = I$$
$$I = \tfrac{P}{E}$$

12. $A = \tfrac{1}{2}h(a+b)$
$$2 \times A = 2 \times \tfrac{1}{2}h(a+b)$$
$$2A = h(a+b)$$
$$\tfrac{2A}{h} = \tfrac{h(a+b)}{h}$$
$$\tfrac{2A}{h} = a+b$$
$$\tfrac{2A}{h} - b = a + b - b$$
$$\tfrac{2A}{h} - b = a$$
$$a = \tfrac{2A}{h} - b$$

14. $P = 2l + 2w$
$$P - 2w = 2l + 2w - 2w$$
$$P - 2w = 2l$$
$$\tfrac{P-2w}{2} = \tfrac{2l}{2}$$
$$\tfrac{P-2w}{2} = l$$
$$l = \tfrac{P-2w}{2}$$

Exercise 7.8

2. $4x^5 \times 18z^9 = 4 \times 18 \times z^5 \times z^9$
$$= 72z^{5+9}$$
$$= 72z^{14}$$

4. $(5a^3)^2 = 5^2 \times (a^3)^2$
$$= 25a^{3\times 2}$$
$$= 25a^6$$

6. $\left(k^{\frac{1}{3}}m^2\right) \times \left(k^{-3}m^{-2}\right)$

$= k^{\frac{1}{3}} \times k^{-3} \times m^2 m^{-2}$

$= k^{\frac{1}{3}+(-3)} \times m^{2+(-2)}$

$= k^{\frac{-9}{3}+\frac{1}{3}} \times m^0$

$= k^{\frac{-8}{3}} \times 1$

$= k^{\frac{-8}{3}}$

End of Chapter Test

2.a. Let the number be $x$
"3 less than" means "subtract 3".
Therefore, 3 less than a number is written as $x - 3$

b. Let the four numbers be $a,\ b,\ c,$ and $d$
"The sum of " means "add".
Therefore, the sum of four numbers is written as $a + b + c + d$

c. An even number is $2x$ (because it can be divided by 2)
Another even number is $2y$
"Sum of" means "add".
Therefore, the sum of two even numbers is written as $2x + 2y$

d. Let one number be $m$.
The next (consecutive) numbers $m + 1$
"Product" means multiply.
Therefore, the product of two consecutive numbers is written as
$m \times (m + 1)$

e. Let the number be $x$

20% means $\frac{20}{100} = \frac{1}{5}$

Therefore, 20% of a number is written as $\frac{1}{5}x$

f. Let one number be $x$
Let the other number be $y$
"ratio" means "divide".

Therefore, the ratio of two numbers is written as $\frac{x}{y}$

4. a. $8x^3 \times 12x^5$

$= 8 \times 12 \times x^3 \times x^5$
$= 96x^{3+5}$
$= 96x^8$

b. $\sqrt{81x^8}$

$= \sqrt{81} \times \sqrt{x^8}$

$= \pm 9x^4$

c. $\frac{15x^{\frac{1}{3}}}{x^{-3}}$

$= 15 \times \left(x^{\frac{1}{3}} \div x^{-3}\right)$

$= 15 \times x^{\frac{1}{3}-(-3)}$

$= 15x^{\frac{1}{3}+\frac{9}{3}}$

$= 15x^{\frac{10}{3}}$

d. $\frac{2x^5y^3 \times 18x^{-5}y^{-3}}{x^{\frac{1}{3}}y^{\frac{1}{4}}}$

$= \frac{2 \times 18 \times x^5 \times x^{-5} \times y^3 \times y^{-3}}{x^{\frac{1}{3}}y^{\frac{1}{4}}}$

$= \frac{36x^{5-5}y^{3-3}}{x^{\frac{1}{3}}y^{\frac{1}{4}}}$

$= \frac{36x^0y^0}{x^{\frac{1}{3}}y^{\frac{1}{4}}}$

$= \frac{36}{x^{\frac{1}{3}}y^{\frac{1}{4}}}$

$= 36x^{\frac{-1}{3}}y^{\frac{-1}{4}}$

# CHAPTER 8: Geometry

Exercise 8.2

2.

6 rays: $\overrightarrow{AB}$ , $\overrightarrow{BA}$ , $\overrightarrow{AC}$ , $\overrightarrow{CA}$ , $\overrightarrow{BC}$ , $\overrightarrow{CB}$

Note that the ray $\overrightarrow{AB}$ is not the same as the ray $\overrightarrow{BA}$

4. C, E, and F are collinear; B, E, and A are co-linear. These points are collinear because, for each set of points, one line can be drawn through all points.

Exercise 8.3

2.b. The angel is obtuse because it larger than a right-angle.

c. Angle size is 150°

d. Supplement = 180° - 150°

= 30°

e. The angle does not have a complement because it is greater than 90°.

4.b. The angle is acute.

c. Angle size is 10°

d. Supplement = 180° - 10°

= 170°

e. Complement = 90° - 10°

= 80°

6. b. The angle is a right-angle

c. 90°

d. Supplement = 180° - 90°

= 90°

e. The angle does not have a complement.

Exercise 8.4

2. The line $\overline{AE}$ is a transversal of parallel lines $\overline{AB}$ and $\overline{DE}$

This means that ∠BAC = ∠DAE

The line $\overline{BD}$ is a transversal of parallel line $\overline{AB}$ and $\overline{DE}$

This means that ∠ABC = ∠CDE

∠ACB = ∠DCE because they are vertically opposite.

4. ∠EPA + ∠ APS + ∠SPQ + ∠QPE
= 360 °

We know that :
∠SRQ = ∠ESD
= ∠EPA
∠RQP = ∠APS
∠PSR =∠QPE

Therefore,

∠QPS +∠PSR +∠SRQ +∠RQP
= ∠QPS +∠QPE + ∠EPA +∠APS
= 360°

Therefore, the sum of the internal angles in a parallelogram is 360°.

Exercise 8.5

2. a. Perimeter
= 8cm+8cm+8cm+8cm+8cm+8cm
= 48 cm

b. Perimeter
$= 5\text{ m} + 5\text{ m} + 5\text{ m} + 5\text{ m}$
$= 20\text{ m}$

c. Perimeter
$= 5\text{ m} + 5\text{ m} + 5\text{ m} + 5\text{ m}$
$= 20\text{ m}$

d. Perimeter
$= 5\text{ mm} + 5\text{ mm} + 6\text{ mm}$
$= 16\text{ mm}$

e. Perimeter
$= 2.5\text{ m} + 1.5\text{ m} + 2\text{m} + 3\text{m} + 2.5\text{ m} + 3.5\text{ m} + 3\text{ m} + 3.5\text{ m} + 3\text{ m} + 3\text{ m} + 4\text{ m} + 12\text{ m} + 15\text{ m}$
$= 58.5\text{ m}$

Exercise 8.6.1

2.a. $2\text{ cm} < 2\text{ cm} + 2\text{ cm}$

c. $12\text{ m} < 14\text{ m} + 6\text{ m}$
$14\text{ m} < 12\text{ m} + 6\text{ m}$
$6\text{ m} < 12\text{ m} + 14\text{ m}$

e. $18\text{ m} < 14\text{ m} + 14\text{ m}$
$14\text{ m} < 18\text{ m} + 14\text{ m}$

4. Area of each triangle
$= \frac{1}{2} \times 6 \times 4\text{ cm}^2$
$= 3 \times 4\text{ cm}^2$
$= 12\text{ cm}^2$

Total area $= 5 \times 12\text{ cm}^2$
$= 60\text{ cm}^2$

Exercise 8.6.2

2. The 2 triangles are congruent by SAS.

The point known as the centre bisects any line that passes through it.

4. Label the 4 corners as A, B, C, and D, going clockwise and starting at the upper left corner of the rectangle.

Label the point in the centre as E.

The opposite sides of rectangle are parallel, because the angle at the corners are all 90°.

Therefore, $\angle CAB = \angle DCA$

$\angle CDB = \angle DBA$

$\angle DEC = \angle AEB$

Therefore , $\triangle ABE = \triangle DEC$

Therefore, $\overline{AE} = \overline{EC}$ and $\overline{BE} = \overline{ED}$
The diagonals bisect each other.

Exercise 8.7

2. a. Area of dining room
$= 4.2\text{ m} \times 3.5\text{ m}$
$= 14.7\text{ m}^2$

Cost of cork tiles $= \$5.6 \times 14.7$
$= \$82.32$

b. Area of hallway $= 8\text{ m} \times 2\text{ m}$
$= 16\text{ m}^2$
Area of living room $= 5\text{ m} \times 8\text{ m}$
$= 40\text{ m}^2$

Total area to be carpeted
$= 40\text{ m}^2 + 16\text{m}^2$
$= 56\text{ m}^2$

Cost of carpet $= \$33 \times 56$
$=\$1848$

c. Area of living room walls:

Two walls are 5 m × 3 m
Two walls are 8 m × 3 m

Area of walls (including any windows or doors)
= 5 m × 3 m + 5 m × 3 m + 8 m × 3 m + 8m × 3m
= $15 m^2 + 15 m^2 + 24 m^2 + 24 m^2$
= $78 m^2$

Amount of paint needed = $\frac{78}{25}$ = 3.12 L

But, since there is at least one door, and there are probably windows, then I would buy 3 L of paint.

Cost of the paint = \$8 × 3
= \$24

Total cost of the renovations
= \$82.32 + \$1848 + \$24
= \$1954.32

4. a. 1 sheet of paper
= 21.5 cm × 28 cm
= $602 cm^2$

$1 m^2$ = 100 cm × 100 cm
= $10000 cm^2$

Number of sheets per square metre
= 10000 ÷ 602
= 16.6 sheets

b. 16.6 sheets have a mass of 15 grams

1 sheet has a mass of 15 grams ÷ 16.6
= 0.904 grams

Number of sheets required to make 28 grams = 28 ÷ 0.904
=30.97

31 sheets of paper make 28 grams.

6. a. The dimensions of the fenced area are 45 m by 30 . (Remember, 10m are added *all of the way around the pool. Draw a diagram to verify these new dimensions.*)

Perimeter of the fenced area
= 45 m +30 m + 45 m + 30 m
= 150 m

b. Fenced area = 45 m × 30 m
= $1350 m^2$

Area of the pool = 25 m × 10 m
= $250 m^2$

Paved area = Fenced area - pool area
= $1350 m^2 - 250 m^2$
= $1100 m^2$

c. Cost of the fencing = \$5 × 150
= \$750

Cost of the tiles = \$15 × 1100
=\$16500

Total cost = \$750 + \$16500
= \$17250

8. a. Area of the rectangle
= 10 m × 20 m
=$200 m^2$

Perimeter of the rectangle
= 10 m +20 m +10 m + 20 m
= 60 m

Ratio of area to perimeter = $\frac{200}{60}$
= $3.\overline{3}$

b. Area of the square = 15 m × 15 m
= $225 m^2$

Perimeter of the square
= 15 m + 15 m + 15 m + 15 m
= 60 m

Ratio of area to perimeter = $\frac{225}{60}$
= 3.75.

c. You would use a square, because it will give you a greater area for a given length of fencing (perimeter).

Exercise 8.8

2. a. An hexagon has 6 sides.

Sum of the internal angles
$= (6 - 2) \times 180°$
$= 720°$

There are also 6 vertices, so each angle
$= 720° \div 6$
$=120°$

b. An octagon has 8 sides.

Sum of the internal angels
$= (8 - 2) \times 180°$
$= 1080°$

There are also 8 vertices, so each angle
$= 1080° \div 8$
$= 135°$

c. A quadrilateral has 4 sides.

Sum of the internal angles
$= (4 - 2) \times 180°$
$= 360°$

There are also 4 vertices, so each angle
$= 360° \div 6$
$= 90°$

Exercise 8.9

2. $C = 2\pi r$

$1 = 2\pi r$
$1 \approx 6.2832r$
$r \approx \frac{1}{6.2832}$
$\approx 0.1592$

$r$ is approximately 0.1592 m or 15.92 cm.

4. Radius of the small pizza
$= 30 \text{ cm} \div 2$
$= 15 \text{ cm}$

Area of the small pizza $= \pi r^2 \text{cm}^2$
$= \pi \times 15^2 \text{ cm}^2$
$= \pi \times 225 \text{ cm}^2$
$\approx 706.86 \text{ cm}^2$

Area of two small pizzas
$\approx 706.86 \text{ cm}^2 \times 2$
$=1413.72 \text{ cm}^2$

Radius of the large pizza $= 45 \text{ cm} \div 2$
$= 22.5 \text{ cm}$

Area of the large pizza $= \pi r^2 \text{cm}^2$
$= \pi \times 22.5^5 \text{cm}^2$
$= \pi \times 506.25 \text{ cm}^2$
$\approx 1590.43 \text{ cm}^2$

The one large pizza is the better value because it has the larger area.

6. Length of Jupiter's orbit is the circumference of the circle that defines its orbit:

$C = 2\pi r$
$= 2 \times \pi \times 778000000 \text{ km}$
$= 2 \times \pi \times 7.78 \times 10^8$
$\approx 4.888 \times 10^{9 \text{ km}}$

Average speed of Jupiter
= distance travelled $\div$ time taken
$= 4.888 \times 10^9 \text{ km} \div 11.86$ years
$= 4.12 \times 10^8$ km per year

Length of Saturn's orbit:

$C = 2\pi r$
$= 2 \times \pi \times 1425000000 \text{ km}$
$= 2 \times \pi \times 1.425 \times 10^9$
$\approx 8.954 \times 10^9 \text{ km}$

Average speed of Saturn
= distance travelled $\div$ time taken
$= 8.954 \times 10^9 \text{ km} \div 29.46$ years
$= 3.04 \times 10^8$ km per year

Jupiter is the faster planet.

Exercise 8.10

2.
Each can has a diameter of 10 cm because the radius is 5 cm.

Dimensions of the carton: 60 cm by 60 cm by 15 cm. This is because there are 6 cans across each of 6 rows. The height of the cans is given.

Volume of the carton $= l \times b \times h$
$= 60 \text{ cm} \times 60\text{cm} \times 15 \text{ cm}$
$= 54\ 000 \text{ cm}^3$

Volume of each can $= \pi r^2 h$

$= \pi \times 5^2 \times 15 \text{ cm}^3$
$= \pi \times 25 \times 15 \text{ cm}^3$
$= 1178.097 \text{ cm}^3$

There are 36 cans in the carton, so the total volume of the cans is:

$V = 36 \times 1178.097 \text{ cm}^3$
$= 42411.492 \text{ cm}^3$

Wasted space
= Volume of the carton - volume of the cans
$= 54\ 000 \text{ cm}^3 - 42\ 411.492 \text{ cm}^3$
$= 11\ 588.508 \text{ cm}^3$

Wasted space as a percentage of total space
$= \frac{11588.508}{54000} \times \frac{100}{1}\%$
$= 21.5\%$

Approximately 21.5% of the carton is wasted.

4. 1 m = 100 cm

1 cubic metre $= 1 \text{ m} \times 1 \text{ m} \times 1 \text{ m}$
$= 100 \text{ cm} \times 100 \text{ cm} \times 100 \text{ cm}$
$= 1\ 000\ 000 \text{ cm}^3$

Since $1 \text{ cm}^3 = 1$ mL,

$1 \text{ m}^3 = 1\ 000\ 000$ mL

1000 mL = 1 L,

so $1 \text{ m}^3 = 1000$ L

6. a. Area of triangular base
$= \frac{1}{2} \times 20 \times 17.32 \text{ mm}^2$
$= 1732 \text{ mm}^2$

Volume of prism
= Area of base × height

$= 173.2 \text{ mm}^2 \times 50 \text{ mm}$
$= 8660 \text{ mm}^3$

b. Area of trapezoidal base
$= \frac{1}{2} \times 5 \times (8+10) \text{ cm}^2$
$= \frac{1}{2} \times 5 \times 18 \text{ cm}^2$
$= 45 \text{ cm}^2$

Volume of prism
= Area of base × height

$= 45 \text{ cm}^2 \times 50 \text{ cm}$
$= 2250 \text{ cm}^3$

End of Chapter Test

2. $C = 2\pi r$

$= 2 \times \pi \times 10 \text{ cm}$

$= 62.83$ cm

4. Area of the square = $s^2$

$= 5^2$ cm$^2$
$= 25$ cm$^2$

Area of the rhombus = $b \times h$

$= 5 \times 4$ cm$^2$
$= 20$ cm$^2$

The square has the greater area.

6. Volume of cylinder = $\pi r^2 h$

$= 1$ L

$\pi r^2 h = 1$ L

$\pi r^2 h = 1000$ cm$^3$

$\pi r^2 = \frac{1000}{15}$

$\pi r^2 \approx 66.6667$

$r^2 \approx \frac{66.6667}{\pi}$

$r^2 \approx 21.2207$

$r \approx 4.61$

The radius of the ice cream container is approximately 4.61 cm.

8. ∠BAC = 75° (because the three angles make a line, and a line is 180° )

∠ABC = 55 ° (corresponding)

∠ACB =50° (corresponding)

10. a. ASA because two angles are marked as having corresponding and equal angles in the other triangle, and the side between the two angles is marked as having a corresponding side of equal length in the other triangle.

b. SSS or SAS because all three sides are marked as having a corresponding side as equal; further, you are given the extra, but unnecessary, piece of information about the equal angles.

c. SAS because the angle between the two sides is marked as equal in both triangles, and the appropriate sides are marked as equal.

# CHAPTER 9: Linear Equations

## Exercise 9.1

2.

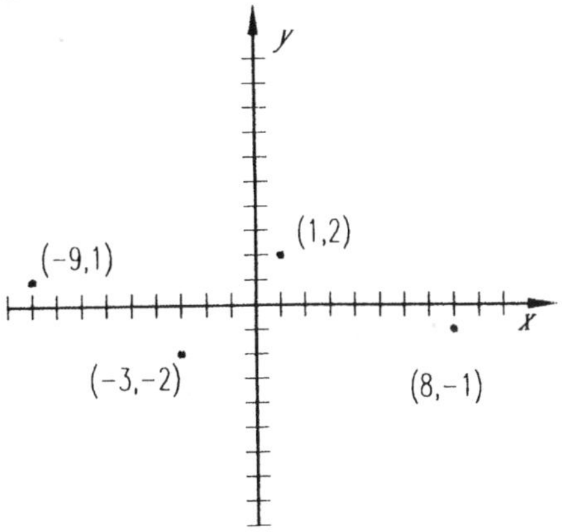


$(1, 2) \Rightarrow$ Quadrant I

$(-9, 1) \Rightarrow$ Quadrant II

$(8, -1) \Rightarrow$ Quadrant IV

$(-3,-2) \Rightarrow$ Quadrant III

## Exercise 9.2

2. We test each point by substituting into the equation, and checking if the LHS is equal to the RHS.

First point $(1, 7)$

LHS

$y$
$= 7$

RHS

$3x + 4$
$= 3(1) + 4$
$= 3 + 4$
$= 7$

LHS = RHS

Since the LHS is equal to the RHS, (1, 7) is a solution to the equation.

Second point (0, 4)

LHS

$y$
$= 4$

RHS

$3x + 4$
$= 3(0) + 4$
$= 0 + 4$

LHS = RHS

Since the LHS is equal to the RHS, (0, 4) is a solution to the equation.

Third point (2, 11)

LHS

$y$
$= 11$
$= 7$

RHS

$3x + 4$
$= 3(2) + 4$
$= 6 + 4$
$= 10$

LHS $\neq$ RHS

Since the LHS is not equal to the RHS, (2, 11) is not a solution to the equation.

4. We test each point by substituting into the equation, and checking if the LHS is equal to the RHS.

First point $(1, 7)$

LHS

$y$
$= 7$

RHS

$5x + 2$
$= 5(1) + 2$
$= 5 + 2$
$= 7$

LHS = RHS

Since the LHS is equal to the RHS, (1, 7) is a solution to the equation.

Second point (0, 2)

LHS

$y$
$= 2$

RHS

$5x + 2$
$= 5(0) + 2$
$= 0 + 2$
$= 2$

LHS = RHS

Since the LHS is equal to the RHS, (0, 2) is a solution to the equation.

Third point (2, 8)

| LHS | RHS |
|---|---|
| $y$ | $5x+2$ |
| $=8$ | $=5(2)+2$ |
| | $=10+2$ |
| | $=12$ |

LHS $\neq$ RHS

Since the LHS is not equal to the RHS, (2, 8) is not a solution to the equation.

6. We test each point by substituting into the equation, and checking if the LHS is equal to the RHS.

First point $(1, 10.5)$

| LHS | RHS |
|---|---|
| $y$ | $10x+0.5$ |
| $=10.5$ | $=10(1)+0.5$ |
| | $=10+0.5$ |
| | $=10.5$ |

LHS $=$ RHS

Since the LHS is equal to the RHS, (1, 10.5) is a solution to the equation.

Second point $(0, 0.5)$

| LHS | RHS |
|---|---|
| $y$ | $10x+0.5$ |
| $=0.5$ | $=10(0)+0.5$ |
| | $=0+0.5$ |
| | $=0.5$ |

LHS $=$ RHS

Since the LHS is equal to the RHS, $(0, 0.5)$ is a solution to the equation.

Third point (2, 8)

| LHS | RHS |
|---|---|
| $y$ | $10x+0.5$ |
| $=8$ | $=10(2)+0.5$ |
| | $=20+0.5$ |
| | $=20.5$ |

LHS $\neq$ RHS

Since the LHS is not equal to the RHS, (2, 8) is not a solution to the equation.

Exercise 9.3

2. $y = -2x$

| $x$ | $y$ | $(x, y)$ |
|---|---|---|
| 1 | $-2$ | $(1, -2)$ |
| 2 | $-4$ | $(2, -4)$ |
| 3 | $-6$ | $(3, -6)$ |

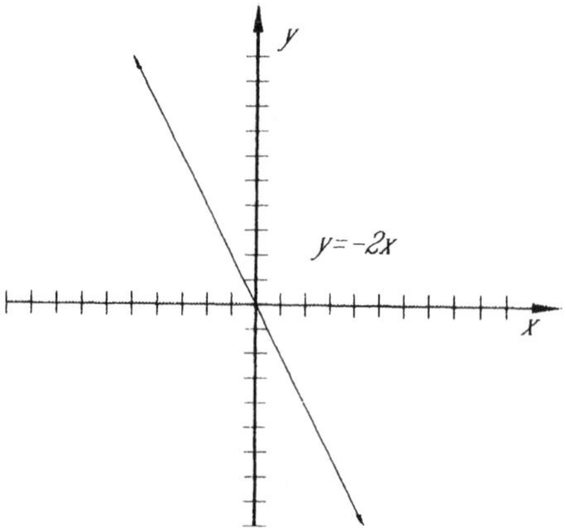


4. $y = \frac{1}{3}x + 5$

| $x$ | $y$ | $(x, y)$ |
|---|---|---|
| 0 | 5 | $(0, 5)$ |
| 3 | 6 | $(3, 6)$ |
| 6 | 7 | $(6, 7)$ |

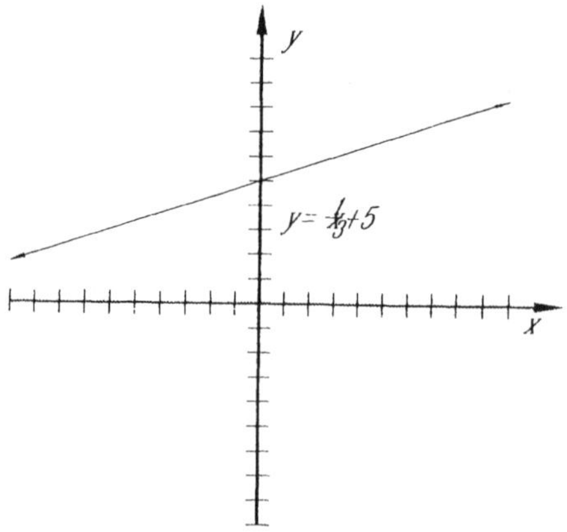

6. $y = 0.4x - 5$

| $x$ | $y$ | $(x, y)$ |
|---|---|---|
| 0 | $-5$ | $(0, -5)$ |
| 1 | $-4.6$ | $(1, -4.6)$ |
| 2 | $-4.2$ | $(2, -4.2)$ |

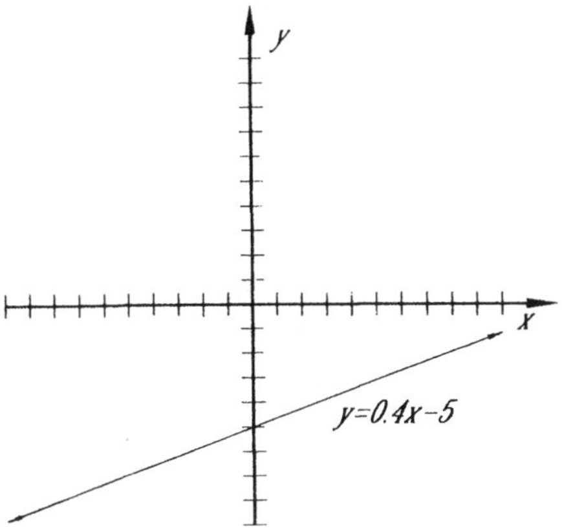


10. $y = 5x - 8$

| $x$ | $y$ | $(x, y)$ |
|---|---|---|
| 0 | $-8$ | $(0, -8)$ |
| 1 | $-3$ | $(1, -3)$ |
| 2 | 2 | $(2, 2)$ |

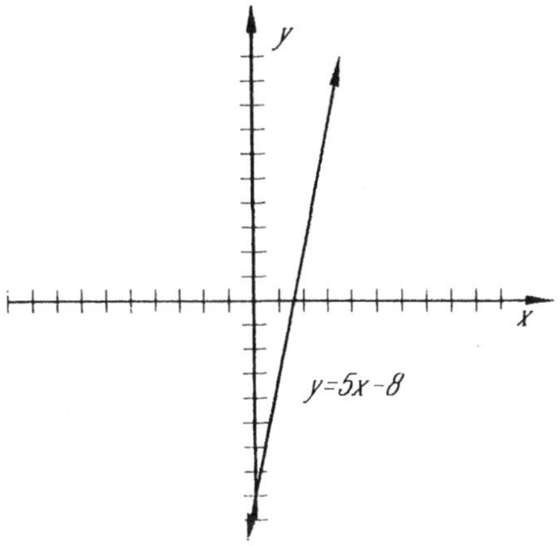


8. $y = 8x + 8$

| $x$ | $y$ | $(x, y)$ |
|---|---|---|
| 0 | 8 | $(0, 8)$ |
| 1 | 16 | $(1, 16)$ |
| $\frac{1}{2}$ | 12 | $(\frac{1}{2}, 12)$ |

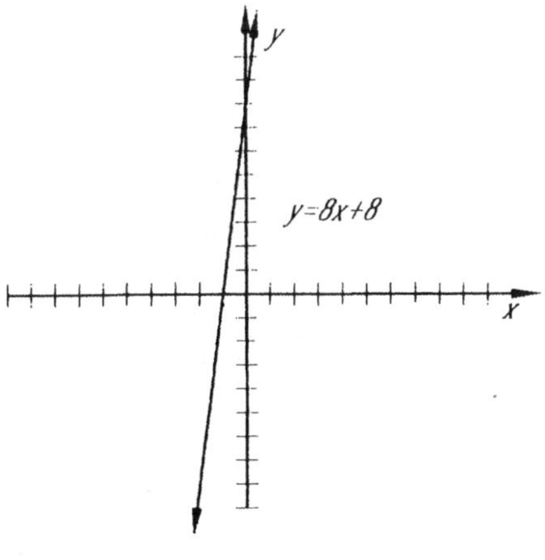


12. $y = -7.5x + 2$

| $x$ | $y$ | $(x, y)$ |
|---|---|---|
| 0 | 2 | $(0, 2)$ |
| 1 | $-5.5$ | $(1, -5.5)$ |
| 2 | $-13$ | $(2, -13)$ |

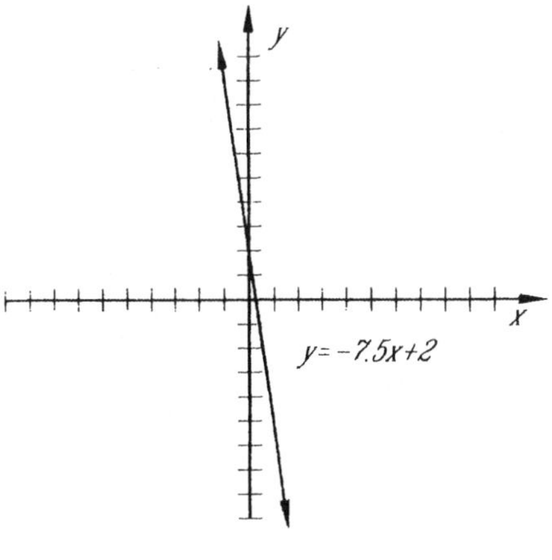

## Exercise 9.4

2. $5x + 7 = 15y$

To find the x-intercept , set y=0.

$$\begin{aligned} 5x + 7 &= 15y \\ 5x + 7 &= 15(0) \\ 5x + 7 &= 0 \\ 5x &= -7 \\ x &= \tfrac{-7}{5} \\ x &= -1.4 \end{aligned}$$

The x-intercept is (-1.4, 0).

To find the y-intercept, set x=0.

$$\begin{aligned} 5x + 7 &= 15y \\ 5(0) + 7 &= 15y \\ 0 + 7 &= 15y \\ 7 &= 15y \\ \tfrac{7}{15} &= y \end{aligned}$$

The y-intercept is (0, $\frac{7}{15}$).

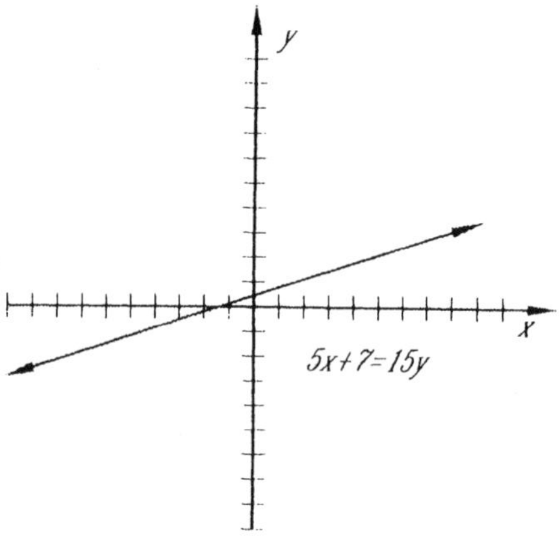


4. $5x + 3y = 15$

To find the x-intercept set y=0.

$$\begin{aligned} 5x + 3y &= 15 \\ 5x + 3(0) &= 15 \\ 5x + 0 &= 15 \\ 5x &= 15 \\ x &= 3 \end{aligned}$$

The x-intercept is (3,0).

To find the y-intercept, set x=0.

$$\begin{aligned} 5x + 3y &= 15 \\ 5(0) + 3y &= 15 \\ 0 + 3y &= 15 \\ 3y &= 15 \\ y &= 5 \end{aligned}$$

The y-intercept is (0,5).

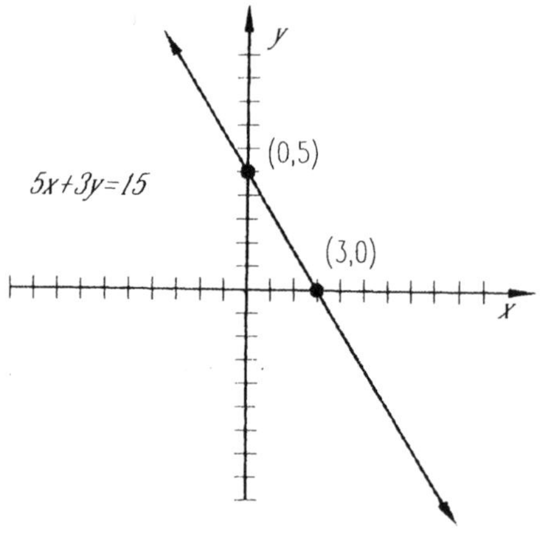

6. $6x - 12 = 6y$

To find the x-intercept, set y=0.

$$
\begin{aligned}
6x - 12 &= 6y \\
6x - 12 &= 6(0) \\
6x - 12 &= 0 \\
6x &= 12 \\
x &= 2
\end{aligned}
$$

The x-intercept is (2,0).

To find the y-intercept, set x=0.

$$
\begin{aligned}
6x - 12 &= 6y \\
6(0) - 12 &= 6y \\
0 - 12 &= 6y \\
-12 &= 6y \\
y &= -2
\end{aligned}
$$

The y-intercept is (0, -2).

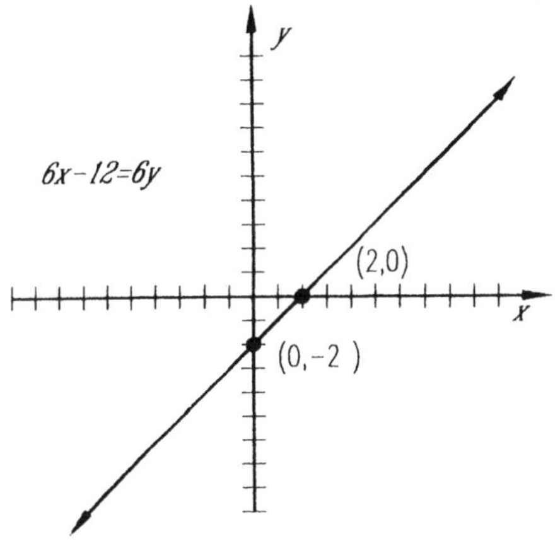


8. $2x + 6y = 18$

To find the x-intercept, set y=0.

$$
\begin{aligned}
2x + 6y &= 18 \\
2x + 6(0) &= 18 \\
2x + 0 &= 18 \\
2x &= 18 \\
x &= 9
\end{aligned}
$$

The x-intercept is (9,0).

To find the y-intercept, set x=0.

$$
\begin{aligned}
2x + 6y &= 18 \\
2(0) + 6y &= 18 \\
0 + 6y &= 18 \\
6y &= 18 \\
y &= 3
\end{aligned}
$$

The y-intercept is (0,3).

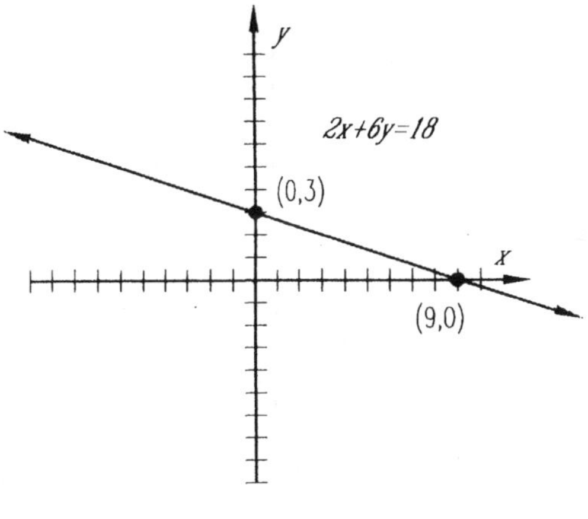

10. $3x - 4y = 12$

To find the x-intercept, set y = 0.

$$\begin{aligned} 3x - 4y &= 12 \\ 3x - 4(0) &= 12 \\ 3x - 0 &= 12 \\ 3x &= 12 \\ x &= 4 \end{aligned}$$

The x-intercept is (4, 0).

To find the y-intercept, set x=0.

$$\begin{aligned} 3x - 4y &= 12 \\ 3(0) - 4y &= 12 \\ 0 - 4y &= 12 \\ -4y &= 12 \\ y &= -3 \end{aligned}$$

The y-intercept is (0, – 3).

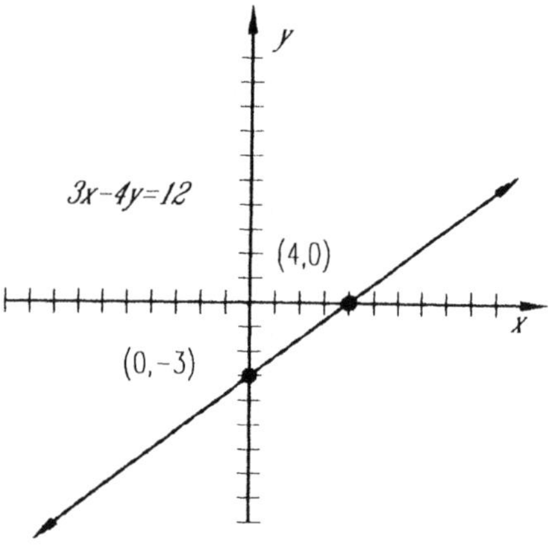


12. $4x - 5 + y = 15x - y + 2$

To find the x-intercept, set y = 0.

$$\begin{aligned} 4x - 5 + y &= 15x - y + 2 \\ 4x - 5 + 0 &= 15x - 0 + 2 \\ 4x - 5 &= 15x + 2 \\ -7 &= 11x \\ -\tfrac{7}{11} &= x \end{aligned}$$

The x-intercept is $(-\frac{7}{11}, 0)$.

To find the y-intercept, set x=0.

$$\begin{aligned} 4x - 5 + y &= 15x - y + 2 \\ 4(0) - 5 + y &= 15(0) - y + 2 \\ -5 + y &= -y + 2 \\ 2y &= 7 \\ y &= \tfrac{7}{2} \end{aligned}$$

The y-intercept is $(0, \frac{7}{2})$.

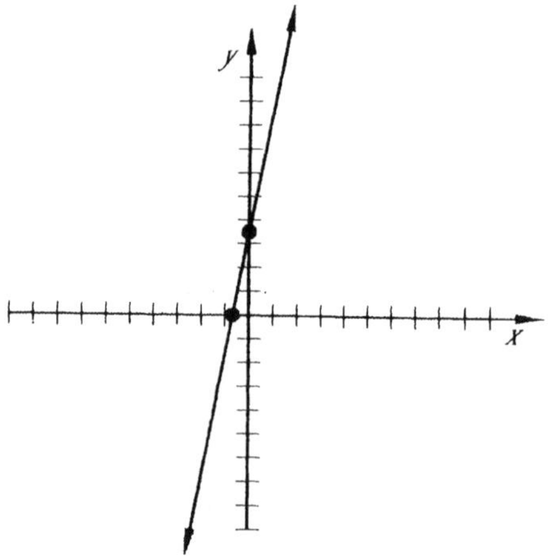

14. $2x - 5 + 2y = 4x - y + 7$

To find the x-intercept, set y = 0.

$$\begin{aligned} 2x - 5 + 2y &= 4x - y + 7 \\ 2x - 5 + 2(0) &= 4x - 0 + 7 \\ 2x - 5 &= 4x + 7 \\ -2x &= 12 \\ x &= -6 \end{aligned}$$

The x-intercept is $(-6, 0)$.

To find the y-intercept, set x=0.

$$\begin{aligned} 2x - 5 + 2y &= 4x - y + 7 \\ 2(0) - 5 + 2y &= 4(0) - y + 7 \\ -5 + 2y &= -y + 7 \\ 3y &= 12 \\ y &= 4 \end{aligned}$$

The y-intercept is (0, 4).

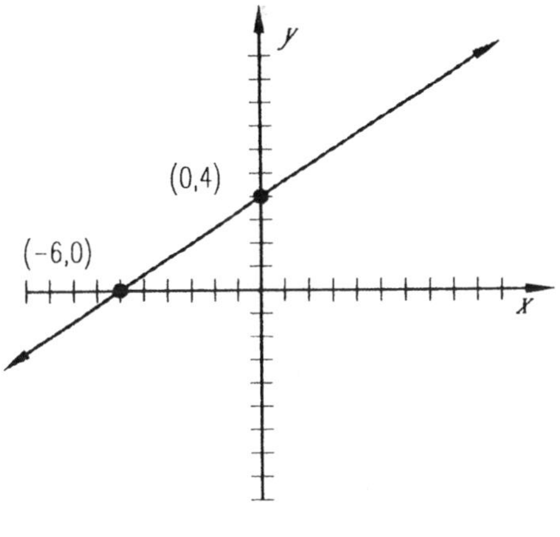


Exercise 9.5

2. slope $= m = \frac{y_1 - y_2}{x_1 - x_2} = \frac{y_2 - y_1}{x_2 - x_1}$

$P_1(1,5)$ $P_2(5,7)$

$$m = \frac{5-7}{1-5} = \frac{-2}{-4} = \frac{1}{2}$$

The slope of the line that contains the points (1, 5) and (5, 7) is $\frac{1}{2}$.

4. slope $= m = \frac{y_1 - y_2}{x_1 - x_2} = \frac{y_2 - y_1}{x_2 - x_1}$

$P_1(-2, 8)$ $P_2(-1, 2)$

$$m = \frac{8-2}{-2-(-1)} = \frac{8-2}{-2+1} = \frac{6}{-1} = -6$$

6. $m = \frac{9-8}{2.5-1.5} = \frac{1}{1} = 1$

8. $m = \frac{-8.2-(-6.2)}{1.3-4} = \frac{-2}{-2.7} = \frac{2}{2.7}$

10. $y = 9x + 4$

$m = 9$ $b = 4$

12. $y = 12x - 4$

$m = 12$ $b = -4$

14. $4x + 5y = 20$

$5y = -4x + 20$

$y = -\frac{4}{5}x + \frac{20}{5}$

$y = -\frac{4}{5}x + 4$

$m = -\frac{4}{5}$ $b = 4$

16. $7y = 12 + 5x$

$y = \frac{5}{7}x + \frac{12}{7}$

$m = \frac{5}{7}$ $b = \frac{12}{7}$

18. $34x + 17y = 17$

$$\begin{aligned} 17y &= -34x + 17 \\ y &= -\frac{34}{17}x + \frac{17}{17} \\ y &= -2x + 1 \end{aligned}$$

$m = -2$ $b = 1$

20. $6y - 2 = 5x - 2y + 3$

$$\begin{aligned} 8y &= 5x + 5 \\ y &= \frac{5}{8}x + \frac{5}{8} \end{aligned}$$

$m = \frac{5}{8}$ $b = \frac{5}{8}$

22. $4y - 3 = 5x - y + 4$

$5y = 5x + 7$

$y = x + \frac{7}{5}$

$m = 1 \qquad b = \frac{7}{5}$

24. $6y = 12x + 3$

$y = 2x + \frac{1}{2}$

$m = 2 \qquad b = \frac{1}{2}$

26. $y = 10x + 4$

$m = 10 \qquad b = 4$

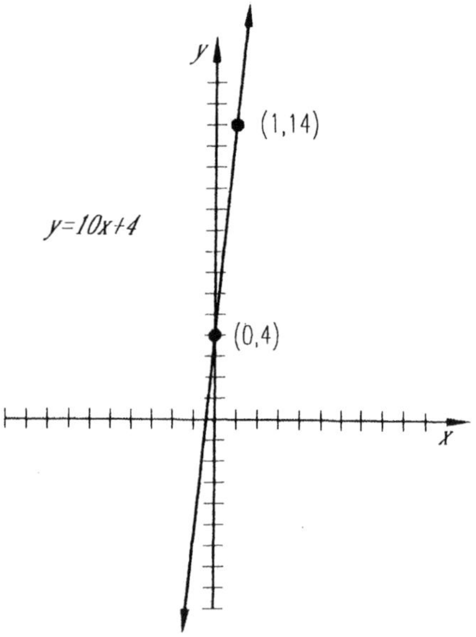


28. $y = -\frac{1}{8}x + 1.5$

$m = -\frac{1}{8} \qquad b = 1.5$

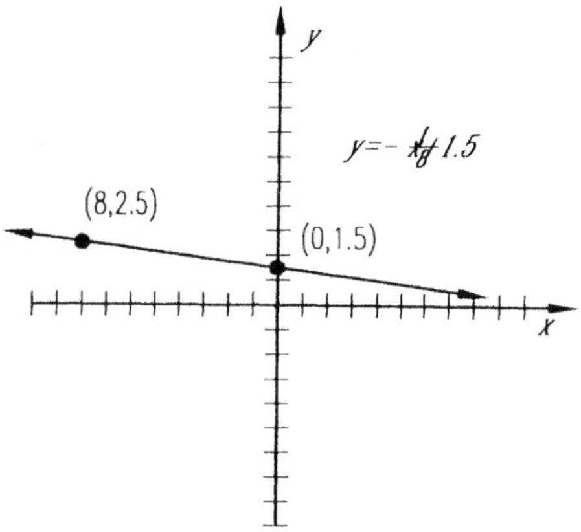


30. $3y = 8x + 3$

$y = \frac{8}{3}x + \frac{3}{3}$

$y = \frac{8}{3}x + 1$

$m = \frac{8}{3} \qquad b = 1$

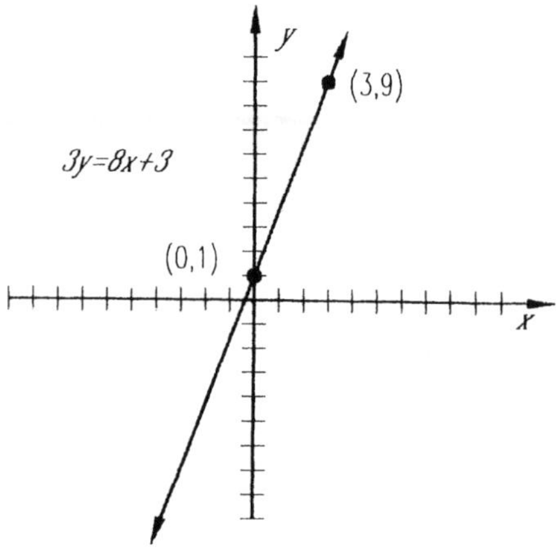


32. $y = -\frac{1}{2}x + 1$

$m = -\frac{1}{2} \qquad b = 1$

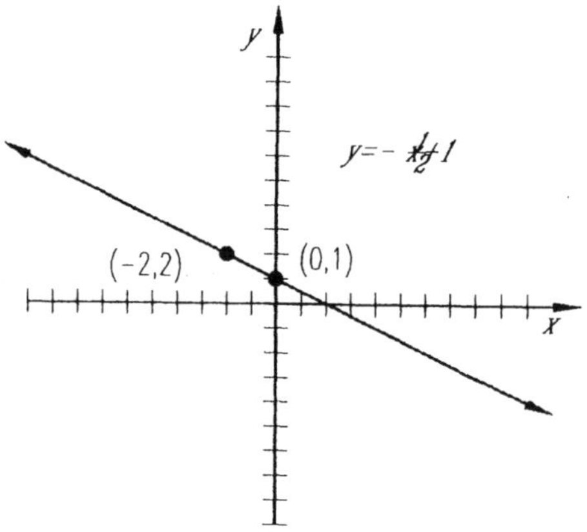

34. $y = -\frac{7}{2}x + 11$

$m = -\frac{7}{2} \qquad b = 11$

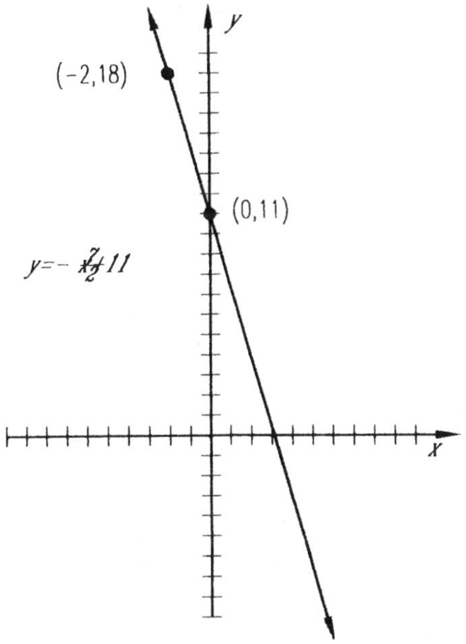


36. $4y = -\frac{1}{2}x + 8$

$y = \frac{-\frac{1}{2}}{4}x + \frac{8}{4}$

$y = -\frac{1}{8}x + 2$

$m = -\frac{1}{8} \qquad b = 2$

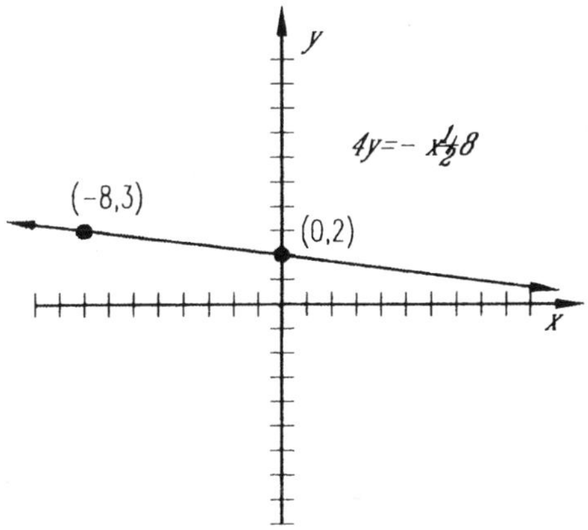


38. $m = -\frac{1}{2} \qquad b = 5$

$y = mx + b$

$y = -\frac{1}{2}x + 5$

40. $m = -12 \qquad b = 14$

$y = -12x + 14$

42. $m = 2 \qquad b = -4$

$y = 2x + (-4)$

$y = 2x - 4$

44. $m = -\frac{1}{3} \qquad b = 5$

$y = -\frac{1}{3}x + 5$

46. $m = -3 \qquad b = 3$

$y = -3x + 3$

48. $m = 2.5 \qquad b = 3.5$

$y = 2.5x + 3.5$

50. $m = 2 \qquad b = 56$

$y = 2x + 56$

Exercise 9.6

2. $m = -5$
$P = (2, 4)$

$y - y_1 = m(x - x_1)$
$y - 4 = -5(x - 2)$
$y = -5x + 14$

4. $m = 0$
$P = (-1, 2)$

$y - y_1 = m(x - x_1)$
$y - 2 = 0(x - (-1))$
$y - 2 = 0$
$y = 2$

6. $m = 2$
$P = (-3, 5)$

$$y - y_1 = m(x - x_1)$$
$$y - 5 = 2(x - (-3))$$
$$y - 5 = 2x + 6$$
$$y = 2x + 11$$

8. $m = 1$
$P = (-5, 2)$

$$y - y_1 = m(x - x_1)$$
$$y - 2 = 1(x - (-5))$$
$$y - 2 = x + 5$$
$$y = x + 7$$

10. $m = -1$
$P = (-1.5, 3.2)$

$$y - y_1 = m(x - x_1)$$
$$y - 3.2 = -1(x - (-1.5))$$
$$y - 3.2 = -x - 1.5$$
$$y = -x + 1.7$$

12. $m = 3$
$P = (-1, 0)$

$$y - y_1 = m(x - x_1)$$
$$y - 0 = 3(x - (-1))$$
$$y = 3x + 3$$

Exercise 9.7

2. $P_1 = (x_1, y_1) = (5, 11)$
$P_2 = (x_2, y2) = (5, 8)$

$$y - y_1 = (\tfrac{y_2 - y_1}{x_2 - x_1})(x - x_1)$$
$$y - 11 = (\tfrac{8-11}{5-5})(x - 5)$$
$$y - 11 = -\tfrac{3}{0}(x - 5)$$

The slope is undefined, therefore x = 5 ; this represents a vertical line.

4. $P_1 = (x_1, y_1) = (-1, -5)$
$P_2 = (x_2, y2) = (0, 6)$

$$y - y_1 = (\tfrac{y_2 - y_1}{x_2 - x_1})(x - x_1)$$
$$y - (-5) = (\tfrac{6-(-5)}{0-(-1)})(x - (-1))$$
$$y + 5 = \tfrac{11}{1}(x + 1)$$
$$y + 5 = 11x + 11$$
$$y = 11x + 6$$

6. $P_1 = (x_1, y_1) = (12, 3)$
$P_2 = (x_2, y2) = (11, 1)$

$$y - y_1 = (\tfrac{y_2 - y_1}{x_2 - x_1})(x - x_1)$$
$$y - 3 = (\tfrac{1-3}{11-12})(x - 12)$$
$$y - 3 = \tfrac{-2}{-1}(x - 12)$$
$$y - 3 = 2(x - 12)$$
$$y - 3 = 2x - 24$$
$$y = 2x - 21$$

8. $P_1 = (x_1, y_1) = (\tfrac{1}{2}, 3)$
$P_2 = (x_2, y2) = (\tfrac{1}{5}, 1)$
$$y - y_1 = (\tfrac{y_2 - y_1}{x_2 - x_1})(x - x_1)$$
$$y - 3 = (\tfrac{1-3}{\frac{1}{5}-\frac{1}{2}})(x - \tfrac{1}{2})$$
$$y - 3 = (\tfrac{-2}{\frac{2-5}{10}})(x - \tfrac{1}{2})$$
$$y - 3 = (\tfrac{-2}{\frac{-3}{10}})(x - \tfrac{1}{2})$$
$$y - 3 = (-2 \times \tfrac{10}{-3})(x - \tfrac{1}{2})$$
$$y - 3 = \tfrac{-20}{-3}(x - \tfrac{1}{2})$$
$$y - 3 = \tfrac{-20}{-3}x - \tfrac{10}{3}$$
$$y = \tfrac{20}{3}x - \tfrac{10}{3} + \tfrac{9}{3}$$
$$y = \tfrac{20}{3}x - \tfrac{1}{3}$$

10. $P_1 = (x_1, y_1) = (\frac{1}{2}, 7)$

$P_2 = (x_2, y2) = (0, 1)$

$y - y_1 = (\frac{y_2 - y_1}{x_2 - x_1})(x - x_1)$

$y - 7 = (\frac{1-7}{0-\frac{1}{2}})(x - \frac{1}{2})$

$y - 7 = \frac{-6}{-\frac{1}{2}}(x - \frac{1}{2})$

$y - 7 = 12(x - \frac{1}{2})$

$y - 7 = 12x - 6$

$y = 12x + 1$

12. $P_1 = (x_1, y_1) = (-\frac{1}{2}, -\frac{1}{3})$

$P_2 = (x_2, y2) = (1, 5)$

$y - y_1 = (\frac{y_2 - y_1}{x_2 - x_1})(x - x_1)$

$y - (-\frac{1}{3}) = \left(\frac{5-(-\frac{1}{3})}{1-(-\frac{1}{2})}\right)(x - (-\frac{1}{2}))$

$y + \frac{1}{3} = \left(\frac{\frac{15}{3}+\frac{1}{3}}{\frac{2}{2}+\frac{1}{2}}\right)(x + \frac{1}{2})$

$y + \frac{1}{3} = \left(\frac{\frac{16}{3}}{\frac{3}{2}}\right)(x + \frac{1}{2})$

$y + \frac{1}{3} = \frac{32}{9}(x + \frac{1}{2})$

$y + \frac{1}{3} = \frac{32}{9}x + \frac{16}{9}$

$y = \frac{32}{9}x + \frac{16}{9} - \frac{1}{3}$

$y = \frac{32}{9}x + \frac{16}{9} - \frac{3}{9}$

$y = \frac{32}{9}x + \frac{13}{9}$

## Exercise 9.8

2. EQ1 $y = 2x + 3$

$m_1 = 2$

EQ2 $4y = -2x + 1$

$y = -\frac{2}{4}x + \frac{1}{4}$

$y = -\frac{1}{2}x + \frac{1}{4}$

$m_2 = -\frac{1}{2}$

If the lines are perpendicular, then $m_1 \times m_2 = -1$.

$2 \times -\frac{1}{2} = -\frac{2}{2} = -1$

since $-1 = -1$

then the lines are perpendicular.

4. EQ1 $y = 2x + 8$

$m_1 = 2$

EQ2 $3y = -7x + 5$

$y = -\frac{7}{3}x + \frac{5}{3}$

$m_2 = -\frac{7}{3}$

If the lines are perpendicular, then $m_1 \times m_2 = -1$.

$2 \times -\frac{7}{3} = -\frac{14}{3}$

since $-\frac{14}{3} \neq -1$

then the lines are not perpendicular.

6. EQ1 $y = \frac{1}{7}x + 8$

$m_1 = \frac{1}{7}$

EQ2 $y = -7x + 5$

$y = -7x + 5$

$m_2 = -7$

If the lines are perpendicular, then $m_1 \times m_2 = -1$.

$\frac{1}{7} \times -7 = -\frac{7}{7} = -1$

since $-1 = -1$

then the lines are perpendicular.

8. EQ1 $y = 6x + 8$

$m_1 = 6$

EQ2 $3y = 18x + 5$

$y = 6x + \frac{5}{3}$

$m_2 = 6$

If the lines are parallel, then $m_1 = m_2$.

$m_1 = m_2$

6=6

The lines are parallel since the slopes are the same.

10. EQ1 $y = 2x + 8$

$m_1 = 2$

EQ2 $3y = -8x + 5$

$y = -\frac{8}{3}x + \frac{5}{3}$

$m_2 = -\frac{8}{3}$

If the lines are parallel, then $m_1 = m_2$.

$m_1 = m_2$

$2 \neq -\frac{8}{3}$

The lines are not parallel since the slopes are not the same.

12. EQ1 $y = 3x + 8$

$m_1 = 3$

EQ2 $9y = 27x + 5$

$y = 3x + \frac{5}{9}$

$m_2 = 3$

If the lines are parallel, then $m_1 = m_2$.

$m_1 = m_2$

$3 = 3$

The lines are parallel since the slopes are the same.

14. $y = 2x + 3$

$slope = m = 2$

$P = (0, 6)$

The equation of the line that is parallel to the equation given, must have the same slope, m = 2.

$y - y_1 = m(x - x_1)$

Point-Slope Equation

$y - 6 = 2(x - 0)$

$y - 6 = 2x - 0$

$y = 2x + 6$

The equation $y = 2x + 6$ is parallel to $y = 2x + 3$ and passes through the point $(0, 6)$.

16. $8y = 5x + 8$

$y = \frac{5}{8}x + 1$

$slope = m = \frac{5}{8}$

$P = (5, 9)$

The equation of the line that is parallel to the equation given, must have the same slope, m = $\frac{5}{8}$.

$y - y_1 = m(x - x_1)$

Point-Slope Equation

$y - 9 = \frac{5}{8}(x - 5)$

$y - 9 = \frac{5}{8}x - \frac{25}{8}$

$y = \frac{5}{8}x - \frac{25}{8} + 9$

$y = \frac{5}{8}x - \frac{25}{8} + \frac{72}{8}$

$y = \frac{5}{8}x + \frac{47}{8}$

The equation $y = \frac{5}{8}x + \frac{47}{8}$ is parallel to $8y = 5x + 8$ and passes through the point (5, 9).

18. $8y = 64x + 2$

$y = 8x + \frac{1}{4}$

$slope = m = 8$

$P = (4, 3)$

The equation of the line that is parallel to the equation given, must have the same slope, m = 8.

$y - y_1 = m(x - x_1)$

Point-Slope Equation

$y - 3 = 8(x - 4)$

$y - 3 = 8x - 32$

$y = 8x - 32 + 3$

$y = 8x - 29$

The equation $y = 8x - 29$ is parallel to $8y = 64x + 2$ and passes through the point $(4, 3)$.

20. $y = 6x + 2$

$slope = m_1 = 6$

$P = (1, 3)$

The equation of the line that is perpendicular to the equation given, must have a slope that satisfies the equation

$m_1 \times m_2 = -1.$

The slope of this line must be $m_2 = -\frac{1}{6}$.

$y - y_1 = m(x - x_1)$

Point-Slope Equation

$y - 3 = -\frac{1}{6}(x - 1)$

$y - 3 = -\frac{1}{6}x + \frac{1}{6}$

$y = -\frac{1}{6}x + \frac{1}{6} + \frac{18}{6}$

$y = -\frac{1}{6}x + \frac{19}{6}$

The equation $y = -\frac{1}{6}x + \frac{19}{6}$ is perpendicular to the equation $y = 6x + 2$ and passes through the point $(1, 3)$.

22. $12y = 6x + 8$

$y = \frac{1}{2}x + \frac{2}{3}$

$slope = m_1 = \frac{1}{2}$

$P = (5, 3)$

The equation of the line that is perpendicular to the equation given, must have a slope that satisfies the equation $m_1 \times m_2 = -1.$
The slope of this line must be $m_2 = -2.$

$y - y_1 = m(x - x_1)$

Point-Slope Equation

$y - 3 = -2(x - 5)$

$y - 3 = -2x + 10$

$y = -2x + 13$

The equation $y = -2x + 13$ is perpendicular to the equation $12y = 6x + 8$ and passes through the point (5, 3).

24. $8y = 24x + 2$

$y = 3x + \frac{1}{4}$

$slope = m_1 = 3$

$P = (0, 3)$

The equation of the line that is perpendicular to the equation given,

must have a slope that satisfies the equation $m_1 \times m_2 = -1$.
The slope of this line must be $m_2 = -\frac{1}{3}$.

$$y - y_1 = m(x - x_1)$$

Point-Slope Equation

$$y - 3 = -\tfrac{1}{3}(x - 0)$$

$$y - 3 = -\tfrac{1}{3}x + 0$$

$$y = -\tfrac{1}{3}x + 3$$

The equation $y = -\frac{1}{3}x + 3$ is perpendicular to the equation $8y = 24x + 2$ and passes through the point (0, 3).

Exercise 9.9.1

2. EQ1 $x - y = -1$

$-y = -x - 1$

$y = x + 1$

$m = 1 \qquad b = 1$

EQ2 $6x + 3y = 12$

$3y = -6x + 12$

$y = -2x + 4$

$m = -2 \qquad b = 4$

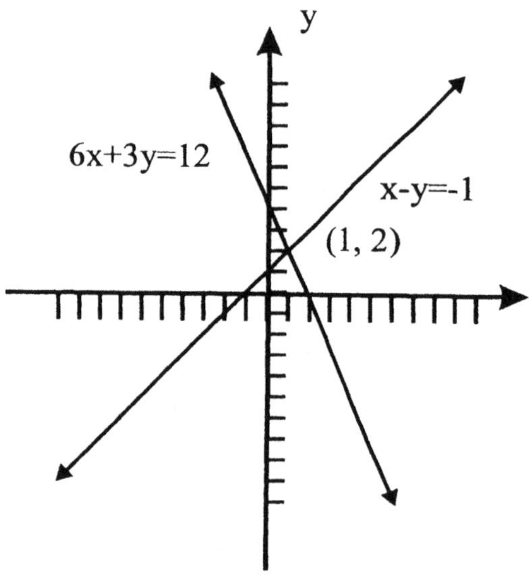


The solution, point of intersection, is (1, 2).

check: $P(1, 2)$

EQ1 $x - y = -1$

| $LHS$ | $RHS$ |
|---|---|
| $x - y$ | $-1$ |
| $= 1 - 2$ | |
| $= -1$ | |

$LHS = RHS$

EQ2 $6x + 3y = 12$

| $LHS$ | $RHS$ |
|---|---|
| $6x + 3y$ | 12 |
| $= 6(1) + 3(2)$ | |
| $= 6 + 6$ | |
| $= 12$ | |

$LHS = RHS$

4. EQ1 $x + 3y = 9$

$3y = -x + 9$

$y = -\frac{x}{3} + 3$

$m = -\frac{1}{3} \qquad b = 3$

EQ2 $6y = 2x + 18$

$y = \frac{1}{3}x + 3$

$m = \frac{1}{3} \qquad b = 3$

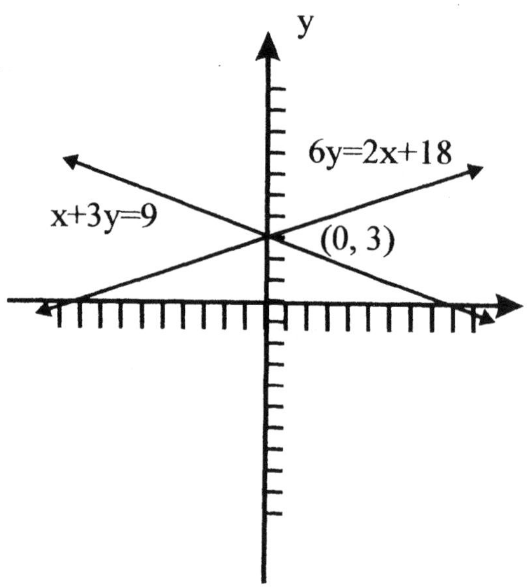


The solution, point of intersection, is (0,3).

check: $P(0, 3)$

EQ1 $x + 3y = 9$

| $LHS$ | $RHS$ |
|---|---|
| $x + 3y$ | $9$ |
| $= 0 + 3(3)$ | |
| $= 9$ | |

$LHS = RHS$

EQ2 $6y = 2x + 18$

| $LHS$ | $RHS$ |
|---|---|
| $6y$ | $2x + 18$ |
| $= 6(3)$ | $= 2(0) + 18$ |
| $= 18$ | $= 18$ |

$LHS = RHS$

6. EQ1 $x - 3y = 9$

$-3y = -x + 9$

$y = \frac{x}{3} + (-3)$

$m = \frac{1}{3}$ $b = -3$

EQ2 $3y = -6x + 12$

$y = -2x + 4$

$m = -2$ $b = 4$

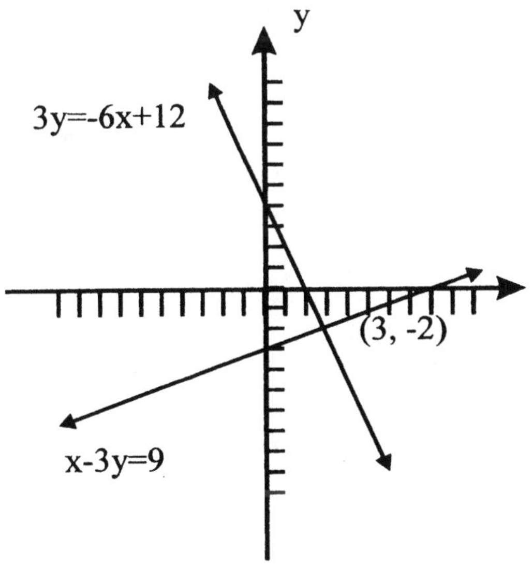


The solution, point of intersection is $(3, -2)$.

check: $P(3, -2)$

EQ1 $x - 3y = 9$

| $LHS$ | $RHS$ |
|---|---|
| $x - 3y$ | $9$ |
| $= 3 - 3(-2)$ | |
| $= 3 + 6$ | |
| $= 9$ | |

$LHS = RHS$

EQ2 $3y = -6x + 12$

| $LHS$ | $RHS$ |
|---|---|
| $3y$ | $-6x + 12$ |
| $= 3(-2)$ | $= -6(3) + 12$ |
| $= -6$ | $= -18 + 12$ |
| | $= -6$ |

$LHS = RHS$

8. EQ1 $2x = 6y$

$y = \frac{1}{3}x$

$m = \frac{1}{3}$ $b = 0$

EQ2 $8y = 4x + 8$

$y = \frac{1}{2}x + 1$

$m = \frac{1}{2}$ $b = 1$

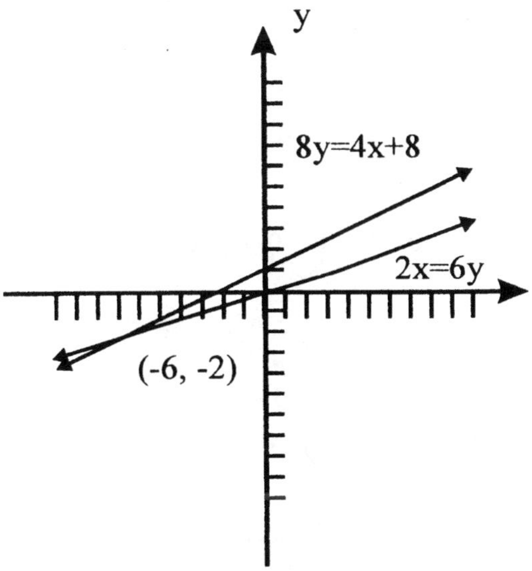


The solution , point of intersection is $(-6, -2)$.

check: $P(-6, -2)$

EQ1 $2x = 6y$

| $LHS$ | $RHS$ |
|---|---|
| $2x$ | $6y$ |
| $= 2(-6)$ | $= 6(-2)$ |
| $= -12$ | $= -12$ |

$LHS = RHS$

EQ2 $8y = 4x + 8$

| $LHS$ | $RHS$ |
|---|---|
| $8y$ | $4x + 8$ |
| $= 8(-2)$ | $= 4(-6) + 8$ |
| $= -16$ | $= -24 + 8$ |
| | $= -16$ |

$LHS = RHS$

10.

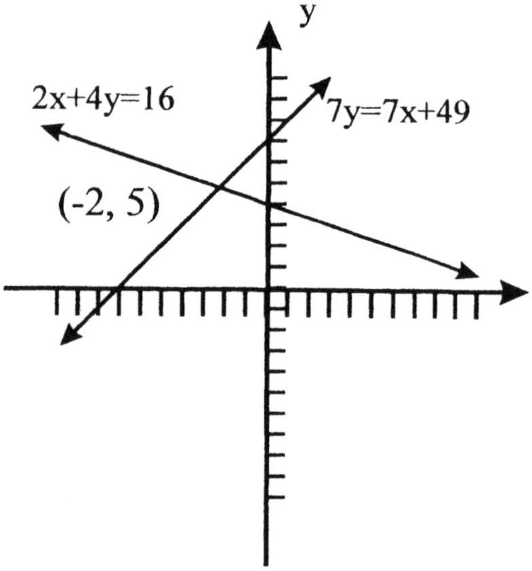


The solution, point of intersection, is (-2, 5).

check: $P(-2, 5)$

EQ1 $x + 4y = 16$

| $LHS$ | $RHS$ |
|---|---|
| $2x + 4y$ | 16 |
| $= 2(-2) + 4(5)$ | |
| $= -4 + 20$ | |
| $= 16$ | |

$LHS = RHS$

EQ2 $7y = 7x + 49$

| $LHS$ | $RHS$ |
|---|---|
| $7y$ | $7x + 49$ |
| $= 7(5)$ | $= 7(-2) + 49$ |
| $= 35$ | $= -14 + 49$ |
| | $= 35$ |

$LHS = RHS$

12.

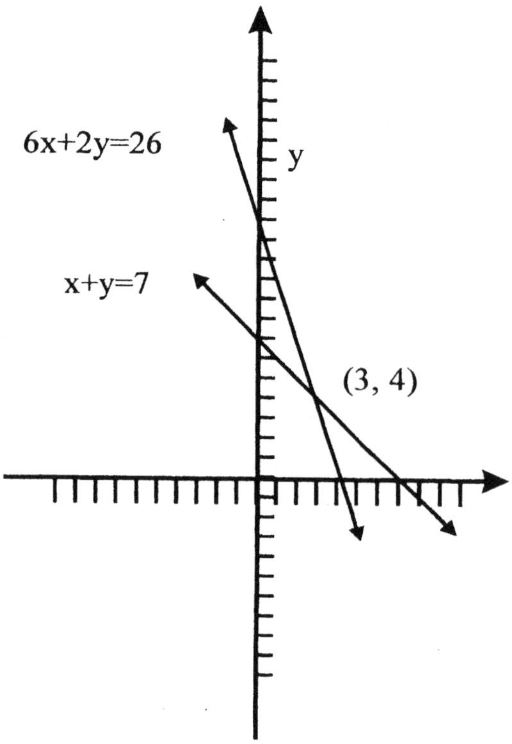


The solution, point of intersection, is (3, 4).

check: $P(3,4)$

EQ1 $6x + 2y = 26$

| $LHS$ | $RHS$ |
|---|---|
| $6x + 2y$ | 26 |
| $= 6(3) + 2(4)$ | |
| $= 18 + 8$ | |
| $= 26$ | |

$LHS = RHS$

EQ2 $x + y = 7$

| $LHS$ | $RHS$ |
|---|---|
| $x + y$<br>$= 3 + 4$<br>$= 7$ | 7 |

$LHS = RHS$

14.

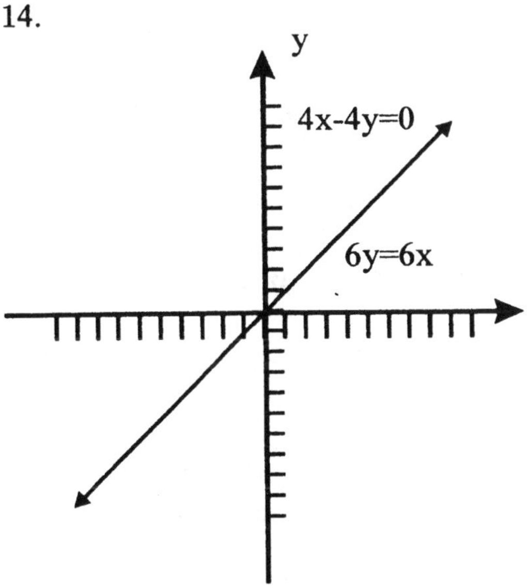


The graphs are the same for both equations. The solution is a line, rather than just a point, therefore any point on the line satisfies the solution of both equations.

Exercise 9.9.2

2. $y = 5x - 11 \quad EQ1$

$2y = 2x + 2 \quad EQ2$

i) Substitute EQ1 into EQ2.

$2y = 2x + 2 \quad EQ2$

$2(5x - 11) = 2x + 2$

$10x - 22 = 2x + 2$

$8x = 24$

$x = 3$

ii) Substitute $x = 3$, into EQ1

$y = 5x - 11 \quad EQ1$

$y = 5(3) - 11$

$y = 15 - 11$

$y = 4$

The solution point is (3, 4).

check: P(3, 4)

$y = 5x - 11 \quad EQ1$

| $LHS$ | $RHS$ |
|---|---|
| $y$<br>$= 4$ | $5x - 11$<br>$= 5(3) - 11$<br>$= 15 - 11$<br>$= 4$ |

$LHS = RHS$

$2y = 2x + 2 \quad EQ2$

| $LHS$ | $RHS$ |
|---|---|
| $2y$<br>$= 2(4)$<br>$= 8$ | $2x + 2$<br>$= 2(3) + 2$<br>$= 6 + 2$<br>$= 8$ |

$LHS = RHS$

The point (3, 4) checks as the solution.

4. $y - x = -4 \quad EQ1$

$2y - x = -1 \, EQ2$

i) Rearrange EQ1 for $y$ :

$y - x = -4 \quad EQ1$

$y = x - 4$

ii) Substitute EQ1 into EQ2.

$2y - x = -1 \quad EQ2$

$2(x - 4) - x = -1$

$2x - 8 - x = -1$

$x = 7$

iii) Substitute $x = 7$, into EQ1

$y - x = -4$ $EQ1$

$y - 7 = -4$

$y = 3$

The solution point is $(7, 3)$.

check: P(7, 3)

$y - x = -4$ $EQ1$

| $LHS$ | $RHS$ |
|---|---|
| $y - x$<br>$= 3 - 7$<br>$= -4$ | $-4$ |

$LHS = RHS$

$2y - x = -1$ $EQ2$

| $LHS$ | $RHS$ |
|---|---|
| $2y - x$<br>$= 2(3) - 7$<br>$= 6 - 7$<br>$= -1$ | $-1$ |

$LHS = RHS$

The point $(7, 3)$ checks as the solution.

6. $y = x - 2$ $EQ1$

$7y = x - 8$ $EQ2$

i) Substitute EQ1 into EQ2.

$7y = x - 8$ $EQ2$
$7(x - 2) = x - 8$
$7x - 14 = x - 8$
$6x = 6$
$x = 1$

ii) Substitute $x = 1$, into EQ1

$y = x - 2$ $EQ1$
$y = 1 - 2$
$y = -1$

The solution point is $(1, -1)$.

check: P(1, − 1)

$y = x - 2$ $EQ1$

| $LHS$ | $RHS$ |
|---|---|
| $y$<br>$= -1$ | $x - 2$<br>$= 1 - 2$<br>$= -1$ |

$LHS = RHS$

$7y = x - 8$ $EQ2$

| $LHS$ | $RHS$ |
|---|---|
| $7y$<br>$= 7(-1)$<br>$= -7$ | $x - 8$<br>$= 1 - 8$<br>$= -7$ |

$LHS = RHS$

The point $(1, -1)$ checks as the solution.

8. $10 - \frac{1}{3}x = y$ $EQ1$

$14x + 2y = -6$ $EQ2$

i) Substitute EQ1 into EQ2.

$14x + 2y = -6$ $EQ2$

$14x + 2(10 - \frac{1}{3}x) = -6$

$14x + 20 - \frac{2}{3}x = -6$

$\frac{42}{3}x - \frac{2}{3}x = -6 - 20$

$\frac{40}{3}x = -26$

$x = -26(\frac{3}{40})$

$x = -\frac{39}{20}$

ii) Substitute $x = -\frac{39}{20}$, into EQ1

$10 - \frac{1}{3}x = y$ $EQ1$

$10 - \frac{1}{3}(-\frac{39}{20}) = y$

$10 - (-\frac{13}{20}) = y$

$\frac{200}{20} + \frac{13}{20} = y$

$\frac{213}{20} = y$

The solution point is $(-\frac{39}{20}, \frac{213}{20})$.

check: P $(-\frac{39}{20}, \frac{213}{20})$.

$10 - \frac{1}{3}x = y$ $EQ1$

$LHS$ $RHS$

$10 - \frac{1}{3}x$ $y$

$= 10 - \frac{1}{3}(-\frac{39}{20})$ $= \frac{213}{20}$

$= 10 + \frac{13}{20}$

$= \frac{200}{20} + \frac{13}{20}$

$= \frac{213}{20}$

$LHS = RHS$

$14x + 2y = -6$ $EQ2$

$LHS$ $RHS$

$14x + 2y$ $-6$

$= 14(-\frac{39}{20}) + 2(\frac{213}{20})$

$= -\frac{546}{20} + \frac{426}{20}$

$= -\frac{120}{20}$

$= -6$

$LHS = RHS$

The point $(-\frac{39}{20}, \frac{213}{20})$ checks as the solution.

Exercise 9.9.3

2. $y - 5x = -11$ $EQ1$
$y - x = 1$ $EQ2$

Multiply EQ2 by $-1$, then add the two equations.

$y - 5x = -11$ $EQ1$

$\underline{-y + x = -1}$ $EQ2 \times -1$

$-4x = -12$

$x = 3$

Substitute $x = 3$ into EQ1.

$y - 5x = -11$
$y - 5(3) = -11$
$y - 15 = -11$
$y = 4$

The solution point is (3, 4).

check: P(3,4)

$y - 5x = -11$ $EQ1$

LHS RHS

$y - 5x$ $-11$
$= 4 - 5(3)$
$= 4 - 15$
$= -11$

$LHS = RHS$

$y - x = 1$ $EQ2$

$LHS$ $RHS$

$y - x$ $1$
$= 4 - 3$
$= 1$

$LHS = RHS$

The point (3, 4) checks and is correct.

4. $y - x = -4$ $EQ1$
$4y - 2x = -2$ $EQ2$

Multiply EQ1 by $-2$, then add the two equations.

$-2y + 2x = 8$ $EQ1 \times -2$

$\underline{4y - 2x = -2}$ $EQ2$

$2y + 0 = 6$

$2y = 6$

$y = 3$

Substitute $y = 3$ into EQ1.

$y - x = -4$ $EQ1$
$(3) - x = -4$
$-x = -4 - 3$
$-x = -7$
$x = 7$

The solution point is (7, 3).

check: P(7, 3)

$y - x = -4 \quad EQ1$

LHS RHS

$y - x$ $\quad -4$
$= 3 - 7$
$= -4$

$LHS = RHS$

$4y - 2x = -2 \quad EQ2$

$LHS \quad RHS$

$4y - 2x$ $\quad -2$
$= 4(3) - 2(7)$
$= 12 - 14$
$= -2$

$LHS = RHS$

The point (7, 3) checks and is correct.

6. $3y - 3x = -6 \quad EQ1$
$7y - x = -8 \quad EQ2$

Multiply EQ2 by $-3$, then add the two equations.

$$\begin{aligned} 3y - 3x &= -6 \quad EQ1 \\ -21y + 3x &= 24 \quad EQ2 \times -3 \\ \hline -18y + 0 &= 18 \\ -18y &= 18 \\ y &= -1 \end{aligned}$$

Substitute $y = -1$ into EQ1.

$$\begin{aligned} 3y - 3x &= -6 \quad EQ1 \\ 3(-1) - 3x &= -6 \\ -3 - 3x &= -6 \\ -3x &= -3 \\ x &= 1 \end{aligned}$$

The solution point is $(1, -1)$.

check: $P(1, -1)$.

$3y - 3x = -6 \quad EQ1$

LHS RHS

$3y - 3x$ $\quad -6$
$= 3(-1) - 3(1)$
$= -3 - 3$
$= -6$

$LHS = RHS$

$7y - x = -8 \quad EQ2$

$LHS \quad RHS$

$7y - x$ $\quad -8$
$= 7(-1) - 1$
$= -7 - 1$
$= -8$

$LHS = RHS$

The point $(1, -1)$ checks and is correct.

8. $10x + 6y = 38 \quad EQ1$

$4x - 10y = 22 \quad EQ2$

Multiply EQ1 by 10, and EQ2 by 6, then add the two equations.

$$\begin{aligned} 100x + 60y &= 380 \quad EQ1 \times 10 \\ 24x - 60y &= 132 \quad EQ2 \times 6 \\ \hline 124x + 0 &= 512 \\ 124x &= 512 \\ x &= \tfrac{512}{124} = \tfrac{128}{31} \end{aligned}$$

Substitute $x = \frac{128}{31}$ into EQ1.

$$\begin{aligned} 10x + 6y &= 38 \quad EQ1 \\ 10(\tfrac{128}{31}) + 6y &= 38 \\ \tfrac{1280}{31} + 6y &= 38 \\ 6y &= \tfrac{1178}{31} - \tfrac{1280}{31} \\ 6y &= -\tfrac{102}{31} \\ y &= -\tfrac{102}{186} = -\tfrac{17}{31} \end{aligned}$$

The solution point is $(\frac{128}{31}, -\frac{17}{31})$.

check: $P(\frac{128}{31}, -\frac{17}{31})$.

$10x + 6y = 38$ $EQ1$

LHS RHS

$10x + 6y$ 38

$= 10(\frac{128}{31}) + 6(-\frac{17}{31})$

$= \frac{1280}{31} - \frac{102}{31}$

$= \frac{1178}{31}$

$= 38$

$LHS = RHS$

$4x - 10y = 22$ $EQ2$

$LHS$ $RHS$

$4x - 10y$ 22

$= 4(\frac{128}{31}) - 10(-\frac{17}{31})$

$= \frac{512}{31} + \frac{170}{31}$

$= \frac{682}{31}$

$= 22$

$LHS = RHS$

The point $P(\frac{128}{31}, -\frac{17}{31})$ checks and is correct.

10. $6x - 4y = 2$ $EQ1$

$-3x + 2y = -1$ $EQ2$

Multiply EQ2 by 2, and then add the two equations.

$$\begin{array}{rl} 6x - 4y = 2 & EQ1 \\ \underline{-6x + 4y = -2} & EQ2 \times 2 \\ 0 + 0 = 0 & \\ 0 = 0 & \end{array}$$

We note that 0 = 0, therefore all solutions for EQ1 will be solutions for EQ2. We also note that EQ1 and EQ2 are the same equations.

12. $6x - 10y = -4$ $EQ1$

$-15x + 15y = 21$ $EQ2$

Multiply EQ1 by 3, and EQ2 by 2, then add the two equations.

$$\begin{array}{rl} 18x - 30y = -12 & EQ1 \times 3 \\ \underline{-30x + 30y = 42} & EQ2 \times 2 \\ -12x + 0 = 30 & \\ -12x = 30 & \\ x = -2.5 & \end{array}$$

Substitute $x = -2.5$ into EQ1.

$$\begin{array}{rl} 6x - 10y = -4 & EQ1 \\ 6(-2.5) - 10y = -4 & \\ -15 - 10y = -4 & \\ -10y = 11 & \\ y = -\frac{11}{10} & \\ y = -1.1 & \end{array}$$

The solution point is $(-2.5, -1.1)$.

check: $P(-2.5, -1.1)$

$6x - 10y = -4$ $EQ1$

LHS RHS

$6x - 10y$ $-4$

$= 6(-2.5) - 10(-1.1)$

$= -15 + 11$

$= -4$

$LHS = RHS$

$-15x + 15y = 21$ $EQ2$

$LHS$ $RHS$

$-15x + 15y$ 21

$= -15(-2.5) + 15(-1.1)$

$= 37.5 - 16.5$

$= 21$

$LHS = RHS$

The point $P(-2.5, -1.1)$ checks and is correct.

14. $\frac{1}{2}x - 7y = -7 \quad EQ1$

$x + \frac{3}{5}y = \frac{6}{10} \quad EQ2$

Multiply EQ1 by 2, and EQ2 by 10, clear the equations of fractions (LCM).

$$x - 14y = -14 \quad EQ1 \times 2$$
$$10x + 6y = 6 \quad EQ2 \times 10$$

Now multiply EQ1 by $-10$, and add the two equations.

$$-10x + 140y = 140 \quad EQ1 \times -10$$
$$\underline{10x + 6y = 6} \quad EQ2$$
$$0 + 146y = 146$$
$$y = 1$$

Substitute $y = 1$ into EQ1.

$$x - 14y = -14 \quad EQ1$$
$$x - 14(1) = -14$$
$$x - 14 = -14$$
$$x = 0$$

The solution point is $(0, 1)$.

check: P$(0, 1)$

$\frac{1}{2}x - 7y = -7$ $EQ1$

| LHS | RHS |
|---|---|
| $\frac{1}{2}x - 7y$ | $-7$ |
| $= \frac{1}{2}(0) - 7(1)$ | |
| $= 0 - 7$ | |
| $= -7$ | |

$LHS = RHS$

$x + \frac{3}{5}y = \frac{6}{10}$ $EQ2$

| $LHS$ | $RHS$ |
|---|---|
| $x + \frac{3}{5}y$ | $\frac{6}{10}$ |
| $= 0 + \frac{3}{5}(1)$ | $= \frac{3}{5}$ |
| $= \frac{3}{5}$ | |

$LHS = RHS$

The point P$(0, 1)$ checks and is correct.

16. $0.2x + 0.4y = 2 \quad EQ1$

$x + y = 6 \quad EQ2$

Multiply EQ1 by 10, to clear the equations of decimals, and multiply EQ2 by -2, then add the two equations.

$$2x + 4y = 20 \quad EQ1 \times 10$$
$$\underline{-2x - 2y = -12} \quad EQ2 \times -2$$
$$0 + 2y = 8$$
$$2y = 8$$
$$y = 4$$

Substitute $y = 4$ into EQ2.

$$x + y = 6 \quad EQ2$$
$$x + 4 = 6$$
$$x = 2$$

The solution point is $(2, 4)$.

check: P$(2, 4)$

$0.2x + 0.4y = 2$ $EQ1$

| LHS | RHS |
|---|---|
| $0.2x + 0.4y$ | 2 |
| $= 0.2(2) - 0.4(4)$ | |
| $= 0.4 + 1.6$ | |
| $= 2$ | |

$LHS = RHS$

$x + y = 6$ $EQ2$

| $LHS$ | $RHS$ |
|---|---|
| $x + y$ | 6 |
| $= 2 + 4$ | |
| $= 6$ | |

$LHS = RHS$

The point P$(2, 4)$ checks and is correct.

18. $\frac{3}{2}x + \frac{3}{2}y = -3 \quad EQ1$

$\frac{2}{5}x - \frac{3}{5}y = \frac{1}{5} \quad EQ2$

Multiply EQ1 by 2, and multiply EQ2 by 5, then add the two equations.

$$\begin{aligned} 3x+3y &= -6 \quad EQ1 \times 2 \\ \underline{2x-3y} &\underline{= 1} \quad EQ2 \times 5 \\ 5x+0 &= -5 \\ 5x &= -5 \\ x &= -1 \end{aligned}$$

Substitute $x = -1$ into EQ2.

$$\begin{aligned} \tfrac{2}{5}x - \tfrac{3}{5}y &= \tfrac{1}{5} \quad EQ2 \\ \tfrac{2}{5}(-1) - \tfrac{3}{5}y &= \tfrac{1}{5} \\ -\tfrac{2}{5} - \tfrac{3}{5}y &= \tfrac{1}{5} \\ -\tfrac{3}{5}y &= \tfrac{3}{5} \\ y &= -1 \end{aligned}$$

The solution point is $(-1, -1)$.

check: $P(-1, -1)$

$\frac{3}{2}x + \frac{3}{2}y = -3 \quad EQ1$

LHS RHS

$\frac{3}{2}x + \frac{3}{2}y$ $\quad -3$

$= \frac{3}{2}(-1) + \frac{3}{2}(-1)$

$= -\frac{3}{2} - \frac{3}{2}$

$= -\frac{6}{2}$

$= -3$

$LHS = RHS$

$\frac{2}{5}x - \frac{3}{5}y = \frac{1}{5} \quad EQ2$

$LHS$ $RHS$

$\frac{2}{5}x - \frac{3}{5}y$ $\quad \frac{1}{5}$

$= \frac{2}{5}(-1) - \frac{3}{5}(-1)$

$= -\frac{2}{5} + \frac{3}{5}$

$= \frac{1}{5}$

$LHS = RHS$

The point $P(-1, -1)$ checks and is correct.

Exercise 9.10

2.

| | A | B | Mix |
|---|---|---|---|
| Amt (l) | A | B | 5l |
| % | 70% | 30% | 50% |
| Amt of Sol. | 0.7(A) | 0.3(B) | $0.5 \times 5 = 2.5$l |

$A + B = 5 \quad$ EQ1

$0.7A + 0.3 = 2.5 \quad$ EQ2

The mechanic should use 2.5 litres of the 70% solution of antifreeze and 2.5 litres of the 30% solution of antifreeze to produce 5 litres of antifreeze containing a 50% solution.

4. $x - y = 21 \quad EQ1$

$x + 3y = 101 \quad EQ2$

The numbers are 20 and 41.

6.

| | d (km) | v ($\frac{Km}{hr}$) | t (hr) |
|---|---|---|---|
| Regina | d | 100 | t+1 |
| Winnipeg | 550 - d | 100 | t |

$100 = \frac{d}{t+1} \quad EQ1$

$100 = \frac{550-d}{t} \quad EQ2$

The trains will meet 3.25 hours, after the first train leaves Regina.

8.

| | Mocha | Blue | Mix |
|---|---|---|---|
| Cost ($\frac{\$}{Kg}$) | \$5 | \$40 | \$20 |
| Amt. of beans (Kg) | x | y | 10 |
| Value ($\frac{\$}{Kg}$)(Kg) | 5x | 40y | $\$200 = 20 \times 10$ |

$5\frac{5}{7}$ kilograms of Mocha beans and $4\frac{2}{7}$ kilograms of Blue mountain beans should be mixed to make the 10 kilogram bag of the blend.

10. $x + y = 25 \quad EQ1$
$2x - 3y = 0 \quad EQ2$

The two numbers are 10 and 15.

12.

| | \$ | Yield | Time | Interest |
|---|---|---|---|---|
| 1st part | x | 7% | 1 yr | 0.07x |
| 2nd part | y | 9.5% | 1yr | 0.095y |
| total | \$5000 | | | \$400 |

$x + y = 5000 \quad EQ1$
$0.07x + 0.095y = 400 \quad EQ2$

\$3000 was invested at a yield of 7% and \$2000 was invested at yield of 9.5%, for a total yield of \$400.

14.

| | Sol. 1 | Sol. 2 | Mix |
|---|---|---|---|
| Amt. sol (l) | x | y | 5 |
| % of sol. | 30% | 60% | 40% |
| Amt. of acid (l) | 0.3x | 0.6y | 2=0.4 × 5 |

$x + y = 5 \quad EQ1$
$0.3x + 0.6y = 2 \quad EQ2$

5 litres of a 40% solution of acid can be made with $3\frac{1}{3}$ litres of the 30% solution acid, and $1\frac{2}{3}$ litres of the 60% solution of acid.

16.

| | v($\frac{km}{hr}$) | d(km) | t(hr) |
|---|---|---|---|
| Army | 60 | d | t + 1 |
| Enemy | 30 | 300 - d | t |

$60 = \frac{d}{t+1} \quad EQ1$

$30 = \frac{300-d}{t} \quad EQ2$

The two armies will meet $3\frac{2}{3}$hr after the army travels east.

Summary Test Chapter 9

2. $(1, 2) \Rightarrow$ Quadrant I
$(-6, 1) \Rightarrow$ Quadrant II
$(2,-1) \Rightarrow$ Quadrant IV
$(-3,-3) \Rightarrow$ Quadrant III

4. We test each point by substituting into the equation, and checking if the LHS is equal to the RHS.

First point $(1, 7)$

| LHS | RHS |
|---|---|
| $y$<br>$= 7$ | $5x + 2$<br>$= 5(1) + 2$<br>$= 5 + 2$<br>$= 7$ |

LHS = RHS

Since the LHS is equal to the RHS, (1, 7) is a solution to the equation.

Second point (0,4)

| LHS | RHS |
|---|---|
| $y$<br>$= 4$ | $5x + 2$<br>$= 5(0) + 2$<br>$= 2$ |

LHS $\neq$ RHS

Since the LHS is not equal to the RHS, (0,4) is not a solution to the equation.

Third point (2, 12)

| LHS | RHS |
|---|---|
| $y$<br>$= 12$ | $5x + 2$<br>$= 5(2) + 2$<br>$= 10+2$<br>$=12$ |

LHS = RHS

Since the LHS is equal to the RHS, (2, 12) is a solution to the equation.

6. $y = 7x + 5$

| $x$ | $y$ | $(x, y)$ |
|---|---|---|
| 0 | 5 | $(0, 5)$ |
| $-1$ | $-2$ | $(-1, -2)$ |
| 1 | 12 | $(1, 12)$ |

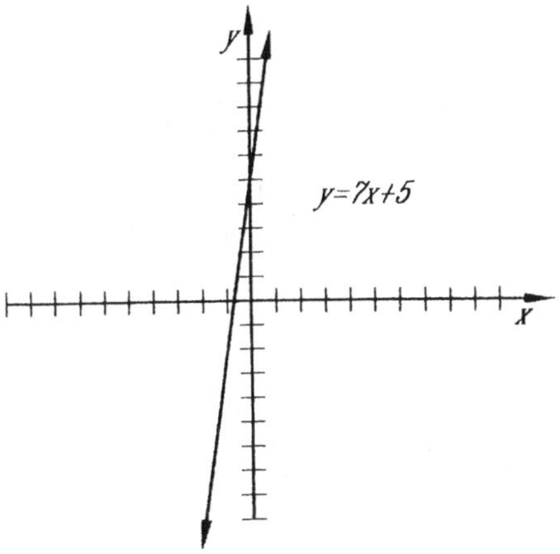


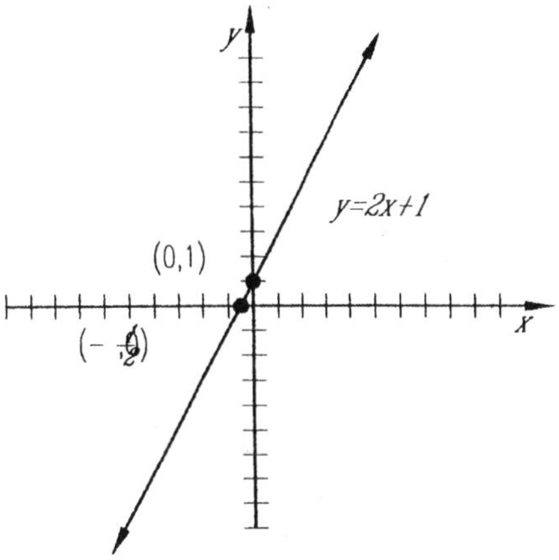


8. $y = 2x + 1$

To find the $x$-intercept , set $y = 0$.

$$y = 2x + 1$$
$$0 = 2x + 1$$
$$-1 = 2x$$
$$-\tfrac{1}{2} = x$$

The x-intercept is $(-\frac{1}{2}, 0)$.
To find the y-intercept, set $x = 0$.

$$y = 2x + 1$$
$$y = 2(0) + 1$$
$$y = 0 + 1$$
$$y = 1$$

The y-intercept is (0, 1).

10. slope $= m = \frac{y_1 - y_2}{x_1 - x_2} = \frac{y_2 - y_1}{x_2 - x_1}$

$P_1(2, 6)$ $P_2(1, 4)$

$$m = \tfrac{4-6}{1-2} = \tfrac{-2}{-1} = 2$$

The slope of the line that contains the points (2,6) and (1, 4) is 2.

12. $m = 5$ $b = 8$

$$y = mx + b$$
$$y = 5x + 8$$

14. $P_1 = (x_1, y_1) = (1, 3)$
$P_2 = (x_2, y2) = (5, 2)$

$$y - y_1 = (\tfrac{y_2 - y_1}{x_2 - x_1})(x - x_1)$$
$$y - 3 = (\tfrac{2-3}{5-1})(x - 1)$$
$$y - 3 = -\tfrac{1}{4}(x - 1)$$
$$y - 3 = -\tfrac{1}{4}x + \tfrac{1}{4}$$
$$y = -\tfrac{1}{4}x + \tfrac{1}{4} + \tfrac{12}{4}$$
$$y = -\tfrac{1}{4}x + \tfrac{13}{4}$$

16. $y = 3x + 4$

$m = 3 \qquad P(1, 2)$

$$y - y_1 = m(x - x_1)$$
$$y - 2 = 3(x - 1)$$
$$y = 3x - 1$$

18. $y = 2x + 6 \quad EQ1$
$y = x+8 \quad EQ2$

$P(2, 10)$

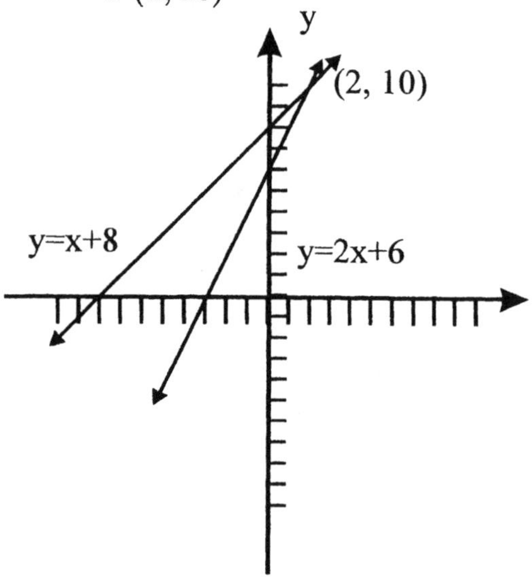


20. $y = 5 - 2x \quad EQ1$
$2x - y = 6 \quad EQ2$

Substitute EQ1 into EQ2.

$$2x - y = 6 \quad EQ2$$
$$2x - (5 - 2x) = 6$$
$$2x - 5 + 2x = 6$$
$$x = \frac{11}{4}$$

Substitute $x = \frac{11}{4}$ into EQ1.

$$y = 5 - 2x \quad EQ1$$
$$y = 5 - 2(\tfrac{11}{4})$$
$$y = 5 - \frac{11}{2}$$
$$y = \frac{10}{2} - \frac{11}{2}$$
$$y = -\frac{1}{2}$$

The point $(\frac{11}{4}, -\frac{1}{2})$ is the solution point.

22. $2x + 3y = 20$ EQ1
$x + 3y = 16 \quad EQ2$

Multiply EQ2 by -1, and add the two equations.

$$2x + 3y = 20$$
$$\underline{-x - 3y = -16}$$
$$x + 0 = 4$$
$$x = 4$$

Substitute x=4 into EQ1.

$2x + 3y = 20 \ EQ1$
$2(4) + 3y = 20$
$8 + 3y = 20$
$y = 4$

The solution point is $P(4, 4)$ and checks.

24.

| | Prin.. | Yield | Time | Int. |
|---|---|---|---|---|
| 1st part | x | 0.10 | 1 yr | 0.1y |
| 2nd part | y | 0.07 | 1 yr | 0.07x |
| Total | $1000 | | | $88 |

$$x + y = 1000 \qquad EQ1$$
$$0.1x + 0.07y = 88 \qquad EQ2$$

$600 was invested at 10% and $400 was invested at 7% for a total yield of $88.

26.

| | $10 | $5 | Total |
|---|---|---|---|
| Bills | $x$ | $y$ | 30 |
| Value | 10x | 5y | $250 |

$x + y = 30 \qquad EQ1$
$10x + 5y = 250 \qquad EQ2$

There are 20 ten dollar bills and 10 five dollar bills for a total of $250.

# CHAPTER 10: Sets and Inequalities

Exercise10.1.1

2. $4y + 7 > 6y - 9$

Substitute $y = 9$

$$\begin{aligned} &4y + 7 > 6y - 9 \\ &4(9) + 7 > 6(9) - 9 \\ &36 + 7 > 54 - 9 \\ &43 > 45 \end{aligned}$$

The statement is false, thus 9 is not a solution.

Substitute $y = 0$

$$\begin{aligned} &4y + 7 > 6y - 9 \\ &4(0) + 7 > 6(0) - 9 \\ &7 > -9 \end{aligned}$$

The statement is true, thus 0 is a solution.

Substitute $y = 7$

$$\begin{aligned} &4y + 7 > 6y - 9 \\ &4(7) + 7 > 6(7) - 9 \\ &28 + 7 > 42 - 9 \\ &35 > 33 \end{aligned}$$

The statement is true, thus 7 is a solution.

4. $6y + 8 \leq 14y - 7$

Substitute $y = 4$

$$\begin{aligned} &6y + 8 \leq 14y - 7 \\ &6(4) + 8 \leq 14(4) - 7 \\ &24 + 8 \leq 56 - 7 \\ &32 \leq 49 \end{aligned}$$

The statement is true, thus 4 is a solution.

Substitute $y = 8$

$$\begin{aligned} &6y + 8 \leq 14y - 7 \\ &6(8) + 8 \leq 14(8) - 7 \\ &48 + 8 \leq 112 - 7 \\ &56 \leq 105 \end{aligned}$$

The statement is true, thus 8 is a solution.

Substitute $y = -14$

$$\begin{aligned} &6y + 8 \leq 14y - 7 \\ &6(-14) + 8 \leq 14(-14) - 7 \\ &-84 + 8 \leq -196 - 7 \\ &-76 \leq -203 \end{aligned}$$

The statement is false, thus -14 is not a solution.

6. $9v - 4 \geq 2v - 2$

Substitute $v = 0$

$$\begin{aligned} &9v - 4 \geq 2v - 2 \\ &9(0) - 4 \geq 2(0) - 2 \\ &-4 \geq -2 \end{aligned}$$

The statement is false, thus 0 is not a solution.

Substitute $v = 1$

$$\begin{aligned} &9v - 4 \geq 2v - 2 \\ &9(1) - 4 \geq 2(1) - 2 \\ &5 \quad 0 \end{aligned}$$

The statement is true, thus 1 is a solution.

Substitute $v = -3$

$$\begin{aligned} &9v - 4 \geq 2v - 2 \\ &9(-3) - 4 \geq 2(-3) - 2 \\ &-31 \geq -8 \end{aligned}$$

The statement is false, thus $-3$ is not a solution.

8. $2 + x < 5x - 4$

Substitute $x = 2$

$$\begin{aligned} &2 + x < 5x - 4 \\ &2 + (2) < 5(2) - 4 \\ &4 < 6 \end{aligned}$$

The statement is true, thus 2 is a solution.

Substitute $x = 3$

$$2 + x < 5x - 4$$
$$2 + (3) < 5(3) - 4$$
$$5 < 11$$

The statement is true, thus 3 is a solution.

Substitute $x =$ -10

$$2 + x < 5x - 4$$
$$2 + (-10) < 5(-10) - 4$$
$$-8 < -54$$

The statement is false, thus $-10$ is not a solution.

10. $7 + y < 4$
$y < 4 - 7$
$y < -3$

$\{y \mid y < -3\}$

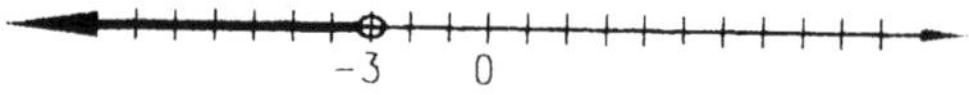


12. $6y - 6 > 8$
$6y > 8 + 6$
$6y > 14$
$y > \frac{14}{6}$
$y > 2\frac{1}{3}$

$\{y \mid y > \frac{14}{6}\}$

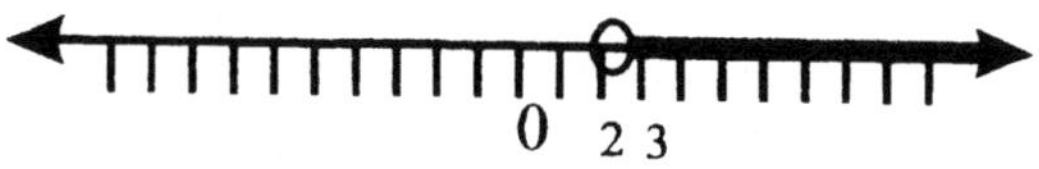


14. $3v - 2 \leq 6$
$3v \leq 6 + 2$
$3v \leq 8$
$v \leq \frac{8}{3}$
$v \leq 2\frac{2}{3}$

$\{v \mid v \leq \frac{8}{3}\}$

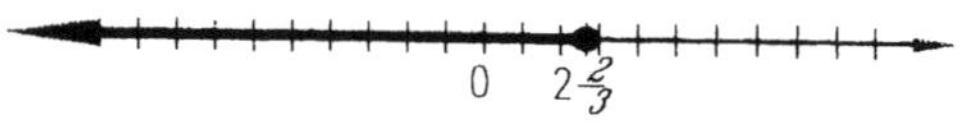


16. $y - 7 > -17$
$y > -10$

$\{y \mid y > -10\}$

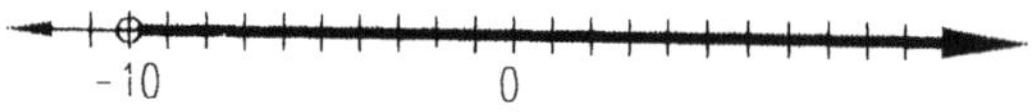


18. $7y < -49$
$y < -7$

$\{x \mid x < -7\}$

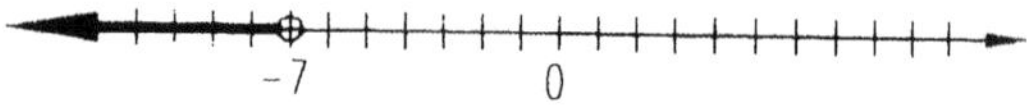


20. $8x < 64$
$x < 8$

$\{x \mid x < 8\}$

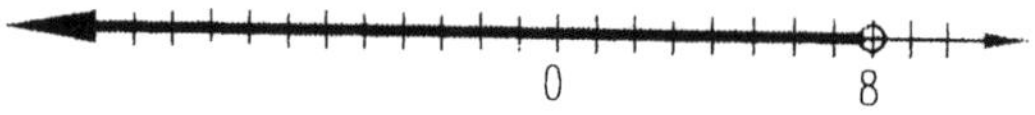


22. $98 + 4y < 12y - 10$
$108 < 8y$
$13\frac{1}{2} < y$

24. $52y > -182$
$y > 3\frac{1}{2}$

26. $45y - 10 \geq y + 60$
$44y \geq 70$
$y \geq \frac{70}{44}$

28. $3(t - 9) + 5 < -2(3t + 2)$
$3t - 27 + 5 < -6t - 4$
$9t < 18$
$t < 2$

30.

$25 - \frac{1}{4}(12x - 4) \geq -(x + 24)$
$25 - 3x + 1 \geq -x - 24$
$-2x > -50$
$x < 25$

32. $4x \geq 64$
$x \geq 16$

34. $4y - 10 \leq y + 50$
$3y \leq 60$
$y \leq 20$

36. $4(2x + 10) - 3x > 4x - 6$
$8x + 40 - 3x > 4x - 6$
$5x + 40 > 4x - 6$
$x > -46$

38. $6(4 - 10x) + 4x < 4(8 + 4x)$
$24 - 60x + 4x < 32 + 16x$
$-56x + 24 < 32 + 16x$
$-72x < 8$
$x > \frac{8}{-72}$
$x > -\frac{1}{9}$

40. $19 - (2x + 3) \leq 2(x + 3) + 2$
$19 - 2x - 3 \leq 2x + 6 + 2$
$-4x \leq -8$
$x \geq 2$

42. $s \leq \$35$

44. $26 - 5x > 12 + x$

46. $4(ab)+5 \leq -27$

48. Case I
$700/week

Case II
$300+0.04(s-3000)

$700<300+0.04s-120$
$700<180+0.04s$
$520<0.04s$
$13000<s$

You would have to sell more than $13000 a week to make more money than the straight salary.

50.

Cost of machine: $2000
Production value of machine: $35/h

Cost of person: $5/h
Production of person:$20/h
Production value of person:
$20-$5/h = $15/h

$35t - 2000 > 15t$
$-2000>-20t$
$100<t$

The machine must operate at least 100 hours to make the purchase of the machine favourable

52. Cost of laser printer:
$\$600 + 0.03p$

Cost of ink jet printer:
$250 +0.16p

$\$600 + 0.03p<\$250 +0.16p$
$350<0.13p$
$\frac{350}{0.13}<p$
$2692.3<p$

You would have to print at least 2693 pages to make the purchase of the laser printer an economical one.

54.

Plan A: \$40/day

Plan B: \$25/day $+ 0.11d$

$40 < 25 + 0.11d$

$136.\overline{36} < d$

To make the unlimited plan economical it would have to travel at least 137 kilometres per day.

56.

Plan A: \$300+\$8t

Plan B:\$10t

$10t > 300 + 8t$
$2t > 300$
$t > 150$

Plan B would be more advantageous for the landscaper over plan A for jobs greater than 150 hours.

58.

First method:
$(E - 3500)0.50$

Second method:
$(E - 7000)0.90$

$(E - 3500)0.50 < (E - 7000)0.90$
$0.5E - 1750 < 0.9E - 6300$
$4550 < 0.4E$
$11375 < E$

The second method would be more economical for loss of earnings greater than \$11375.

Exercise 10.1.2

2. $\{9\}$

4. $\{69\}$

6. $\{5, 6, 41\}$

8. $\emptyset$

10.

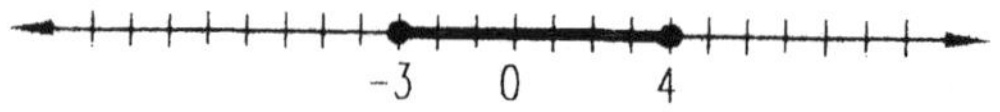


12. $9 > -x \geq -3$
$-9 < x \leq 3$

14.

16. $7 > -v \geq -5$

$7 > -v \Rightarrow -7 < v$

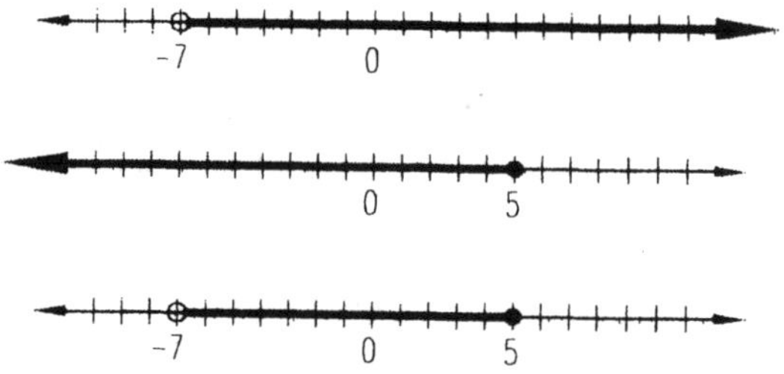


$-v \geq -5 \Rightarrow v \leq 5$

18.

$$\begin{aligned} -10 < 3x - 2 \text{ and } & \quad 3x - 4 \le 10 \\ -8 < 3x & \quad 3x \le 14 \\ -\tfrac{8}{3} < x & \quad x \le \tfrac{14}{3} \end{aligned}$$

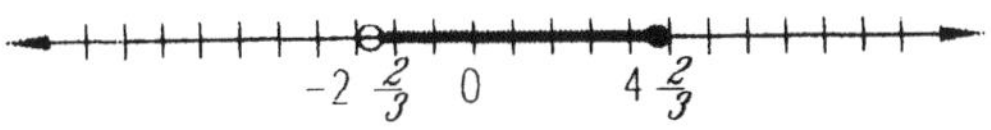


20.

$$\begin{aligned} 5x - 7 < 3 \quad \text{and} & \quad 5 - 6x > 7 \\ 5x < 10 & \quad -6x > 2 \\ x < 2 & \quad x < -\tfrac{1}{3} \end{aligned}$$

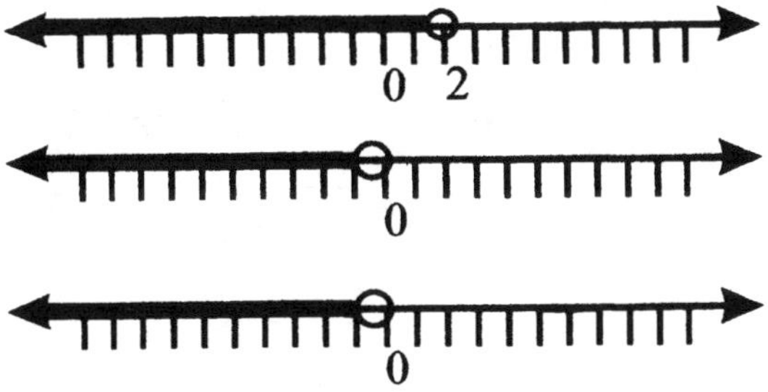


22.

$$-3 < x + 3 < 7$$

$$\begin{aligned} -3 < x + 3 \text{ and } & \quad x + 3 < 7 \\ -6 < x & \quad x < 4 \end{aligned}$$

$$\{x | -6 < x < 4\}$$

24.

$$2 < 3y + 6 \le 9$$

$$\begin{aligned} 2 < 3y + 6 \quad \text{and} & \quad 3y + 6 \le 9 \\ -4 < 3y & \quad 3y \le 3 \\ -\tfrac{4}{3} < y & \quad y \le 1 \end{aligned}$$

$$\{x | -1\tfrac{1}{3} < y \le 1\}$$

26.

$$-7 < x + 3 < 12$$

$$\begin{aligned} -7 < x + 3 \text{ and } & \quad x + 3 < 12 \\ -10 < x & \quad x < 9 \end{aligned}$$

$$\{x | -10 < x < 9\}$$

28.

$$4 < 6y + 3 \le 9$$

$$\begin{aligned} 4 < 6y + 3 \quad \text{and} & \quad 6y + 3 \le 9 \\ 1 < 6y & \quad 6y \le 6 \\ \tfrac{1}{6} < y & \quad y \le 1 \end{aligned}$$

$$\{y | \tfrac{1}{6} < y \le 1\}$$

30. $\{1, 3, 6, 9, 10, 12\}$

32. $\{1, 5, 9\}$

34. $\{1, 2, 3, 6, 8, 9, 11, 13\}$

36. $\{0, 2, 19\}$

38.

$$\begin{aligned} x + 6 < -2 \text{ or } & \quad x + 6 > 4 \\ x < -8 & \quad x > -2 \end{aligned}$$

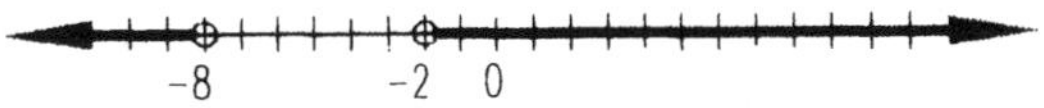


40.

$$\begin{aligned} 2x - 6 \le -3 \text{ or } & \quad x - 9 \ge 3 \\ 2x \le 3 & \quad x \ge 12 \\ x \le \tfrac{3}{2} & \end{aligned}$$

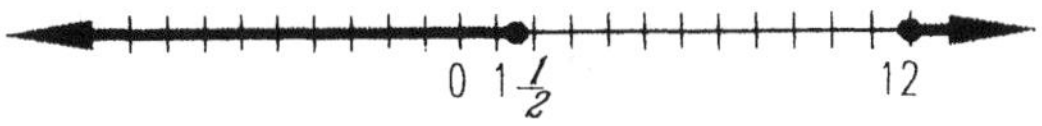


42.

$$\begin{aligned} x + 4 < -2 \text{ or } & \quad x + 8 > 4 \\ x < -6 & \quad x > -4 \end{aligned}$$

$$\{x | x < -6 \text{ or } x > -4\}$$

44.

$$\begin{aligned} 3x - 6 < -3 \text{ or } & \quad x - 6 \ge 3 \\ 3x < 3 & \quad x \ge 9 \\ x \le 1 & \end{aligned}$$

$$\{x | x \le 1 \text{ or } x \ge 9\}$$

Exercise 10.1.3

2.$|9x| = |9|\,|x| = 9|x|$

4.$|-12x^2| = |-12|\,|x^2| = 12x^2$

6.$\left|\frac{12x}{-12x^2}\right| = \left|\frac{1}{-x}\right| = \frac{1}{|x|}$

8.$|13x| = |13||x| = 13|x|$

10.$|-13x^2| = |-13||x^2| = 13x^2$

12. $\left|\frac{22x}{-11x^2}\right| = \left|\frac{2}{-x}\right| = \frac{2}{|x|}$

14.$\{-9, 9\}$

16.$\{0\}$

18.{-19,19}

20.{-7.4,7.4}

22. $|4y + 7| = 12$

$4y + 7 = 12$ *or* $4y + 7 = -12$
$4y = 5$ $\quad 4y = -19$
$y = \frac{5}{4}$ $\quad y = -\frac{19}{4}$

$\{\frac{5}{4}, -\frac{19}{4}\}$

24. $|14x + 12| = |x + 4|$

$14x + 12 = x + 4$
$13x = -8$
$x = -\frac{8}{13}$

*or*

$14x + 12 = -(x + 4)$
$14x + 12 = -x - 4$
$15x = -16$
$x = -\frac{16}{15}$

$\{-\frac{8}{15}, -\frac{16}{15}\}$

26.

$|4x - 4| = 12$

$4x - 4 = 12$ *or* $4x - 4 = -12$
$4x = 16$ $\quad 4x = -8$
$x = 4$ $\quad x = -2$

$\{-2, 4\}$

28.

$|4x + 8| = |x + 4|$

$4x + 8 = x + 4$
$3x = -4$
$x = -\frac{4}{3}$

*or*

$4x + 8 = -(x + 4)$
$4x + 8 = -x - 4$
$5x = -12$
$x = -\frac{12}{5}$

$\{-\frac{4}{3}, -\frac{12}{5}\}$

30.

$|4x| - 3 = 15$
$|4x| = 18$
$4|x| = 18$
$|x| = 4.5$

$\{-4.5, 4.5\}$

32.

$7|x| - 7 = 12$
$7|x| = 19$
$|x| = \frac{19}{7}$

$\{-\frac{19}{7}, \frac{19}{7}\}$

34.

$|2x + 8| = |x + 2|$

$2x + 8 = x + 2$
$x = -6$

*or*

$2x + 8 = -(x + 2)$
$2x + 8 = -x - 2$
$3x = -10$
$x = -\frac{10}{3}$

$\{-6, -\frac{10}{3}\}$

36.

$|8p + 5| = |4p + 3|$

$8p + 5 = 4p + 3$
$4p = -2$
$p = -\frac{1}{2}$

*or*

$8p + 5 = -(4p + 3)$
$8p + 5 = -4p - 3$
$12p = -8$
$p = -\frac{8}{12}$
$p = -\frac{2}{3}$

$\{-\frac{1}{2}, -\frac{2}{3}\}$

38.

$|t - 6| = |6 - t|$

$t - 6 = 6 - t$
$2t = 12$
$t = 6$

or

$t - 6 = -(6 - t)$
$t - 6 = -6 + t$
$0 = 0$

All real numbers are solutions.

40.

$|\frac{2x-3}{3}| = |\frac{7-2x}{7}|$

$\frac{2x-3}{3} = \frac{7-2x}{7}$
$7(2x - 3) = 3(7 - 2x)$
$14x - 21 = 21 - 6x$
$20x = 42$
$x = \frac{42}{20}$
$x = 2.1$

*or*

$\frac{2x-3}{3} = -(\frac{7-2x}{7})$

$7(2x - 3) = 3(-7) + 3(2x)$
$14x - 21 = -21 + 6x$
$8x = 0$
$x = 0$

$\{0, 2.1\}$

42.

$|x - 5| < 2$

$-2 < x - 5 < 2$
$3 < x < 7$

$\{x|3 < x < 7\}$

44.

$|5y - 5| \leq 15$

$-15 \leq 5y - 5 \leq 15$
$-2 \leq y \leq 4$

$\{y| -2 \leq y \leq 4\}$

46.

$|6-4x| \geq 18$

$6-4x \geq 18$
$-4x \geq 12$
$x \leq -3$

*or*

$6-4x \leq -18$
$-4x \leq -24$
$x \geq 6$

$\{x \mid x \geq 6 \text{ or } x \leq -3\}$

48.

$|2x-6| < 2$

$-2 < 2x-6 < 2$
$2 < x < 4$

$\{x|2 < x < 4\}$

50.

$|8x-5| < 6$

$-6 < 8x-5 < 6$
$-\frac{1}{8} < x < \frac{11}{8}$

$\{x| -\frac{1}{8} < x < \frac{11}{8}\}$

52.

$|9x-5| < 23$

$-23 < 9x-5 < 23$
$-2 < x < \frac{28}{9}$

$\{x| -2 < x < \frac{28}{9}\}$

54.

$|3x-3| \geq 9$

$3x-3 \geq 9$
$3x \geq 12$
$x \geq 4$

*or*

$3x-3 \leq -9$
$3x \leq -6$
$x \leq -2$

$\{x|x \leq -2 \text{ or } x \geq 4\}$

56.

$|5x+5| < 30$

$-30 < 5x+5 < 30$
$-7 < x < 5$

$\{x| -7 < x < 5\}$

58.

$|x-5| < 260$

$-260 < x-5 < 260$
$-255 < x < 265$

$\{x| -255 < x < 265\}$

Exercise 10.1.4

2.

$5x+2y > 19$ $\qquad P=(3,4)$

| *LHS* | *RHS* |
|---|---|
| $5x+2y$<br>$=5(3)+2(4)$<br>$=15+8$<br>$=23$ | 19 |

Since 23>19, the point (3,4) is a solution

4.

$3x+4y < 7$

*Graph :*

$3x+4y=7$
$4y = -3x+7$
$y = -\frac{3}{4}x+\frac{7}{4}$

$m = -\frac{3}{4} \qquad b=\frac{7}{4}$

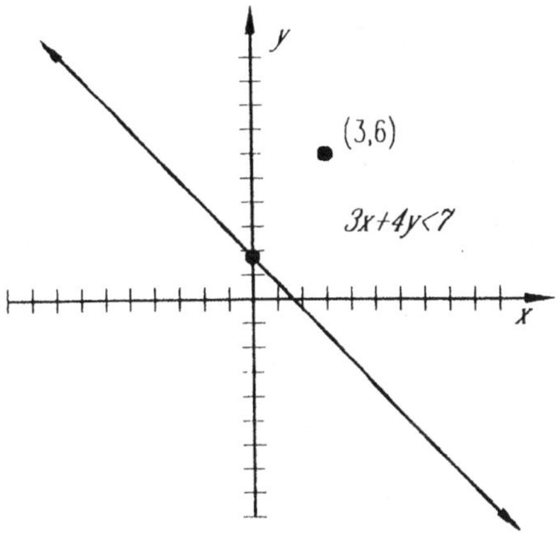


*Test* $P(3,6)$

| *LHS* | *RHS* |
|---|---|
| $3x+4y$ | 7 |
| $=3(3)+4(6)$ | |
| $=9+24$ | |
| $=23$ | |

Since 23<7 is false, all points on the opposite side of our test point are solutions to our inequality.

6.

$8x+y>20$

*Graph* : $8x+y=20$

$y=-8x+20$

$m=-8 \quad b=20$

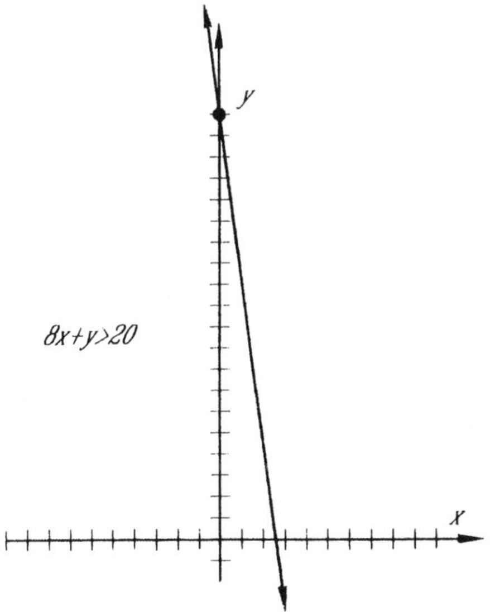


*Test* $P(1,1)$

| *LHS* | *RHS* |
|---|---|
| $8x+y$ | 20 |
| $=8(1)+1$ | |
| $=8+1$ | |
| $=9$ | |

Since 9>20 is false, all points on the opposite side of our test point are solutions to our inequality.

8.

$x+4y<2$

*Graph* : $x+4y=2$

$y=-\frac{1}{4}x+\frac{1}{2}$

$m=-\frac{1}{4} \quad b=\frac{1}{2}$

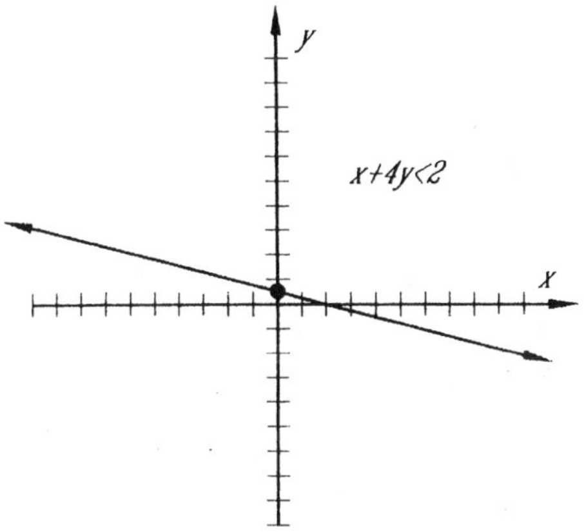


*Test* $P(0,0)$

| *LHS* | *RHS* |
|---|---|
| $x+4y$ | 2 |
| $=0+4(0)$ | |
| $=0$ | |

Since 0<2 is true all points on the same side of our test point are solutions to our inequality.

10.

$7x + 5y < 0$

*Graph* : $7x + 5y = 0$

$y = -\frac{7}{5}x$

$m = -\frac{7}{5}$ $b = 0$

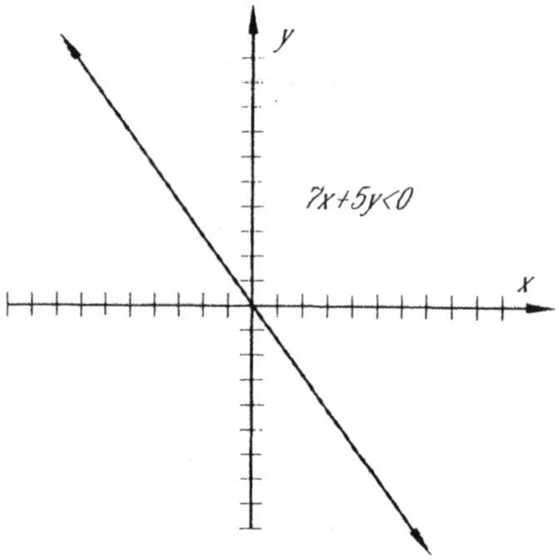


Test $P(1,1)$

| LHS | RHS |
|---|---|
| $7x + 5y$ $= 7(1) + 5(1)$ $= 7 + 5$ $= 12$ | 0 |

Since 12<0 is false, all points on the opposite side of our test point are solutions to our inequality.

Summary Test Chapter10

2. $-6x \geq 72$

$x \leq -12$

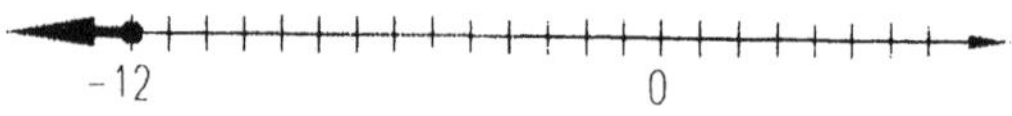


4. $3 + 2(ab) < 53$

6. $\{5,9\}$

8. $\{1, 5, 6, 8, 9, 12, 14, 15\}$

10. $\left|\frac{12x^2}{-4x}\right| = |-3x| = 3|x|$

12. $|4y - 5| = |6y + 12|$

$4y - 5 = 6y + 12$
$-2y = 17$
$y = -8.5$

*or*

$4y + 5 = -(6y + 12)$
$4y - 5 = -6y - 12$
$10y = -7$
$y = -0.7$

$\{-8.5, -0.7\}$

14. $3x + 2y > 18$ $P(1,3)$

$3(1) + 2(3) > 18$
$3 + 6 > 18$
$9 > 18$

Since 9>18 is false, the point (1,3) is not a solution to the inequality.

# CHAPTER 11: Factoring

## Exercise 11.2

2. a. $8 = 2 \times 4$
$= 2 \times 2 \times 2$

b. $195 = 5 \times 39$
$= 5 \times 3 \times 13$

c. $93 = 3 \times 31$

g. $168 = 2 \times 84$
$= 2 \times 2 \times 42$
$= 2 \times 2 \times 2 \times 21$
$= 2 \times 2 \times 2 \times 3 \times 7$

## Exercise 11.3

2.a. $5 = 5 \times 1$

$24 = 2 \times 2 \times 2 \times 3$
$= 2^3 \times 3$

$LCM = 2^3 \times 3 \times 5$
$= 8 \times 3 \times 5$
$= 120$

b. $15 = 5 \times 3$

$18 = 2 \times 9$
$= 2 \times 3 \times 3$
$= 2 \times 3^3$

$LCM = 5 \times 2 \times 3^2$
$= 5 \times 2 \times 9$
$= 90$

c. $7 = 7 \times 1$

$15 = 5 \times 3$

$LCM = 7 \times 5 \times 3$
$= 105$

d. $5 = 5 \times 1$

$s = s \times 1$

$LCM = 5 \times s$
$= 5s$

e. $4w = 2 \times 2 \times w$

$5y = 5 \times y$

$LCM = 2 \times 2 \times w \times 5 \times y$
$= 20wy$

f. $6p = 2 \times 3 \times p$

$12k = 2 \times 2 \times 3 \times k$

$LCM = 12pk$

g. $4 = 2 \times 2$

$5d = 5 \times d$

$LCM = 2 \times 2 \times 5 \times d$
$= 20d$

h. $120 = 2 \times 60$
$= 2 \times 2 \times 30$
$= 2 \times 2 \times 2 \times 15$
$= 2 \times 2 \times 2 \times 3 \times 5$

$63 = 7 \times 9$
$= 7 \times 3 \times 3$

$LCM = 2 \times 2 \times 2 \times 3 \times 3 \times 5 \times 7$
$= 2520$

i. $2wq = 2 \times w \times q$

$5 = 5 \times 1$

$LCM = 2 \times w \times q \times 5$
$= 10wq$

j.
$$\begin{aligned}36\text{kq} &= 2 \times 18 \times k \times q \\ &= 2 \times 2 \times 9 \times k \times q \\ &= 2 \times 2 \times 3 \times 3 \times k \times q\end{aligned}$$

$$\begin{aligned}54k &= 6 \times 9 \times k \\ &= 2 \times 3 \times 3 \times 3 \times k\end{aligned}$$

$$\begin{aligned}LCM &= 2 \times 2 \times 3 \times 3 \times 3 \times k \times q \\ &= 108kq\end{aligned}$$

k.
$$\begin{aligned}18k^2 &= 2 \times 9 \times k \times k \\ &= 2 \times 3 \times 3 \times k \times k\end{aligned}$$

$$k = k \times 1$$

$$\begin{aligned}LCM &= 2 \times 3 \times 3 \times k \times k \\ &= 18k^2\end{aligned}$$

Exercise 11.4

2.
$$\begin{aligned}4 + 12t &= 4 \times 1 + 4 \times 3t \\ &= 4(1 + 3t)\end{aligned}$$

3.
$$\begin{aligned}10s - 5t &= 5 \times 2s - 5 \times t \\ &= 5(2s - t)\end{aligned}$$

6.
$$\begin{aligned}2g - 4gh &= 2g \times 1 - 2g \times 2h \\ &= 2g(1 - 2h)\end{aligned}$$

8.

$$\begin{aligned}st + 5rs + s^2 &= s \times t + s \times 5r + s \times s \\ &= s(t + 5r + s)\end{aligned}$$

Exercise 11.5

2.

$$\begin{aligned}(p + 3)(p + 1) &= p(p + 1) + 3(p + 1) \\ &= p^2 + p + 3p + 3 \\ &= p^2 + 4p + 3\end{aligned}$$

4.
$$\begin{aligned}(3t - 5)(4 - 2t) &= 3t(4 - 2t) - 5(4 - 2t) \\ &= 12t - 6t^2 - 20 + 10t \\ &= -6t^2 + 22t - 20\end{aligned}$$

6.
$$(x + y - z)(x + 2y + 3z)$$

$$= x(x + 2y + 3z) + y(x + 2y + 3z) - z(x + 2y + 3z)$$

$$= x^2 + 2xy + 3xz + xy + 2y^2 + 3yz - xz - 2yz - 3z^2$$

$$= x^2 + 2y^2 - 3z^2 + 2xy + xy + 3xz - xz + 3yz - 2yz$$

$$= x^2 + 3xy + 2xz + 2y^2 + yz - 3z^2$$

Exercise 11.6

2.a. $x^2 - 10x + 24 = 0$

We need two numbers that multiply to give 24 and add to give $-10$ :

$$(-6) \times (-4) = +24$$

$$(-6) + (-4) = -10$$

The two numbers are $-6$ and $-4$.

$$(x - 6)(x - 4) = 0$$

Therefore, $x - 6 = 0$ OR $x - 4 = 0$

$x = 6$ OR 4

b. $x^2 + 10x - 24 = 0$

We need two numbers that multiply to give $-24$ and add to give 10 :

$$(+12) \times (-2) = -24$$

$$(+12) + (-2) = +10$$

The two numbers are $+12$ and $-2$.

$$(x+12)(x-2) = 0$$

Therefore, $x + 12 = 0$ OR $x - 2 = 0$

$x = -12$ OR $+2$

c. $x^2 + 2x + 1 = 0$

We need two numbers that multiply to give 1 and add to give 2 :

$$(1) \times (1) = 1$$

$$(1) + (1) = 2$$

The two numbers are 1 and 1.

$$(x+1)(x+1) = 0$$

Therefore, $x + 1 = 0$ OR $x + 1 = 0$

$x = -1$

## Exercise 11.7

2. In the expression $5x^2 + 9x + 3$, $a = 5$, $b = 9$, and $c = 3$

$$\frac{-b \pm \sqrt{b^2 - 4ac}}{2a} = \frac{-9 \pm \sqrt{9^2 - 4 \times 5 \times 3}}{2 \times 5}$$

$$= \frac{-9 \pm \sqrt{81 - 60}}{10}$$

$$= \frac{-9 \pm \sqrt{21}}{10}$$

$$5x^2 + 9x + 3 = \left(x - \frac{-9+\sqrt{21}}{10}\right)\left(x - \frac{-9-\sqrt{21}}{10}\right)$$

4. In the expression $3x^2 + 4x + 1$, $a = 3$, $b = 4$, and $c = 1$

$$\frac{-b \pm \sqrt{b^2 - 4ac}}{2a} = \frac{-4 \pm \sqrt{4^2 - 4 \times 3 \times 1}}{2 \times 3}$$

$$= \frac{-4 \pm \sqrt{16 - 12}}{6}$$

$$= \frac{-4 \pm \sqrt{4}}{6}$$

$$= \frac{-4 \pm 2}{6}$$

$$= \frac{-4-2}{6} \text{ or } \frac{-4+2}{6}$$

$$= \frac{-6}{6} \text{ or } \frac{-2}{6}$$

$$= -1 \text{ or } \frac{-1}{3}$$

$$3x^2 + 4x + 1 = (x+1)\left(x + \tfrac{1}{3}\right)$$

## Exercise 11.8

2. $x^2 + 18x + 81 = (x+9)^2$

because $9^2 = 81$ and $2 \times 9 = 18$

4. $x^2 - 20x + 100 = (x - 10)^2$

because $(-10)^2 = 100$ and $2 \times (-10) = -20$

## Exercise 11.9

2. $y^2 - 36 = y^2 - 6^2$

$$= (y-6)(y+6)$$

4. $x^4 - 81 = (x^2)^2 - 9^2$

$= (x^2 - 9)(x^2 + 9)$

$= (x^2 - 3^2)(x^2 + 9)$

$= (x - 3)(x + 3)(x^2 + 9)$

6. $9x^2 - 16y^2 = (3x)^2 - (4y)^2$

$= (3x - 4y)(3x + 4y)$

Exercise 11.10

2. $t^3 - 27 = t^3 - 3^3$

$= (t - 3)(t^2 + 3t + 9)$

4. $t^3 - s^3 = (t - s)(t^2 + st + s^2)$

Exercise 11.11

2.
$(r + 2s)(x + y) + (2r - 2s)(x + y)$

$= (x + y)(r + 2s + 2r - 2s)$

$= (x + y)(r + 2r + 2s - 2s)$

$= (x + y)(3r)$

$= 3r(x + y)$

4. $w^2 + 8p - 2w - 16p$

$= w^2 - 2w + 8p - 16p$
$= w^2 - 2w - 8p$

There are no obvious factors.

6. $5xy - 5xw + 5zy - 5zw$

$= 5x(y - w) + 5z(y - w)$

$= (5x + 5z)(y - w)$

End of Chapter Test

2.a. $x^2 - 64 = x^2 - 8^2$
$= (x - 8)(x + 8)$

b. $w^2 + 5w + 6 = (w + 2)(w + 3)$

because $2 \times 3 = 6$ and $2 + 3 = 5$

c. $3x^2 - 3x - 36$
$= 3(x^2 - x - 12)$
$= 3(x - 4)(x + 3)$

because $(-4) \times (+3) = -12$ and $(-4) + (+3) = -1$

d. $s^3 + 8 = s^3 + 2^3$

$= (s + 2)(s^2 - 2s + 4)$

e. $9w^2 - 16p^2 = (3w)^2 - (4p)^2$
$= (3w - 4p)(3w + 4p)$

f. $4t^3 - 16t = 4t(t^2 - 4)$
$= 4t(t^2 - 2^2)$
$= 4t(t - 2)(t + 2)$

g. $27x^3 - 125 = (3x)^3 - 5^3$
$= (3x - 5)(9x^2 - 15x + 25)$

h. $x^4 - 81 = (x^2)^2 - 9^2$
$= (x^2 - 9)(x^2 + 9)$
$= (x - 3)(x + 3)(x^2 + 9)$

i. $x^2 + 15x - 100$
$= (x+20)(x-5)$

because $(20) \times (-5) = -100$ and
$(20) + (-5) = 15$

j. $3x^2 - 11x - 20$
$= (3x+4)(x-5)$

| | |
|---|---|
| $3x$ | $x$ |
| 4 | $-5$ |

k. $x^2 - 9t^2$
$= x^2 - (3t)^2$
$= (x-3t)(x+3t)$

4. $(x+2)^3$

$= (x+2)(x+2)(x+2)$
$= (x+2)(x^2 + 2x + 2x + 4)$
$= (x+2)(x^2 + 4x + 4)$
$= x^3 + 4x^2 + 4x + 2x^2 + 8x + 8$
$= x^3 + 6x^2 + 12x + 8$

which is not equal to $x^3 + 8$

# CHAPTER 12: Quadratic Equation and Graphing

## Exercise 12.1

2. $4v^2 - v = 12$
$4v^2 - v - 12 = 0$

$a = 4 \quad b = -1 \qquad c = -12$

4. $x^2 + 5x + 7 = -6$
$x^2 + 5x + 13 = 0$

$a = 1 \quad b = 5 \quad c = 13$

6. $x^2 + 8 = 12$
$x^2 - 4 = 0$

$a = 1 \quad b = 0 \quad c = -4$

8. $(x-4)^2 = 16$
$x - 4 = \pm\sqrt{16}$
$x = 4 \pm 4$
$x = 4 + 4 \quad or\ x = 4 - 4$
$x = 8 \qquad x = 0$

10. $(x-12)^2 = 5$
$x - 12 = \pm\sqrt{5}$
$x = 12 \pm \sqrt{5}$

$x = 12 + \sqrt{5}\ or\ x = 12 - \sqrt{5}$

12. $(x-9)^2 = 121$
$x - 9 = \pm\sqrt{121}$
$x = 9 \pm 11$
$x = 9 + 11 \quad or\ x = 9 - 11$
$x = 20 \qquad x = -2$

## Exercise 12.2.1

2. $-3x^2 + 12 = 0$
$-3x^2 = -12$
$x^2 = 4$
$x = \pm\sqrt{4}$
$x = \pm 2$

4. $16x^2 + 64 = 0$
$16x^2 = -64$
$x^2 = -\frac{64}{16}$
$x^2 = -4$
$x = \pm\sqrt{-4}$
$x = \pm\sqrt{-1 \times 4}$
$x = \pm\sqrt{-1} \times \sqrt{4}$
$x = \pm 2i$

Note: $\sqrt{-1} = i$

6. $3x^2 - 3 = 0$
$3x^2 = 3$
$x^2 = 1$
$x = \pm\sqrt{1}$
$x = \pm 1$

14. $(x+1)^2 = -16$
$x + 1 = \pm\sqrt{-16}$
$x + 1 = \pm\sqrt{-1} \times \sqrt{16}$
$x + 1 = \pm 4i$
$x = -1 \pm 4i$

16. $(v+1)^2 = 64$
$v + 1 = \pm\sqrt{64}$
$v + 1 = \pm 8$
$v = -1 \pm 8$

$v = 7 \quad or \qquad v = -9$

18. $\left(t + \frac{5}{7}\right)^2 = \frac{15}{49}$
$t + \frac{5}{7} = \pm\sqrt{\frac{15}{49}}$
$t = -\frac{5}{7} \pm \sqrt{\frac{15}{49}}$
$t = -\frac{5}{7} \pm \frac{\sqrt{15}}{7}$

$t = \frac{-5 \pm \sqrt{15}}{7}$

Exercise 12.2.2

2. $x^2 + 2x - 8 = 0$

$x^2 + 2x + 1 = 8 + 1$

$(x+1)^2 = 9$

$x + 1 = \pm\sqrt{9}$

$x = -1 \pm 3$

$x = 2$ *or* $x = -4$

4. $x^2 = 10x - 3$

$x^2 - 10x + 25 = -3 + 25$

$(x-5)^2 = 22$

$x - 5 = \pm\sqrt{22}$

$x = 5 \pm \sqrt{22}$

6. $x^2 - x = 12$

$x^2 - x + \frac{1}{4} = 12 + \frac{1}{4}$

$\left(x - \frac{1}{2}\right)^2 = \frac{49}{4}$

$x - \frac{1}{2} = \pm\sqrt{\frac{49}{4}}$

$x = \frac{1}{2} \pm \frac{7}{2}$

$x = 4$ *or* $x = -3$

8. $x^2 - 8x - 8 = 0$

$x^2 - 8x = 8$

$x^2 - 8x + \left(-\frac{8}{2}\right)^2 = 8 + \left(-\frac{8}{2}\right)^2$

$x - 8x + 16 = 8 + 16$

$(x-4)^2 = 24$

$x - 4 = \pm\sqrt{24}$

$x = 4 \pm 2\sqrt{6}$

10. $x^2 - 6x - 8 = 0$

$x^2 - 6x = 8$

$x^2 - 6x + \left(-\frac{6}{2}\right)^2 = 8 + \left(-\frac{6}{2}\right)^2$

$x^2 - 6x + 9 = 8 + 9$

$(x-3)^2 = 17$

$x - 3 = \pm\sqrt{17}$

$x = 3 \pm \sqrt{17}$

12. $x^2 + 4x - 13 = 0$

$x^2 + 4x = 13$

$x^2 + 4x + \ + 4 = 13 + 4$

$(x+2)^2 = 17$

$x + 2 = \pm\sqrt{17}$

$x = -2 \pm \sqrt{17}$

14. $x^2 + 9x - 9 = 0$

$x^2 + 9x = 9$

$x^2 + 9x + \left(\frac{9}{2}\right)^2 = 9 + \left(\frac{9}{2}\right)^2$

$x^2 + 9x + \left(\frac{9}{2}\right)^2 = \frac{36}{4} + \frac{81}{4}$

$(x + \frac{9}{2})^2 = \frac{117}{4}$

$x + \frac{9}{2} = \pm\sqrt{\frac{117}{4}}$

$x = \frac{-9 \pm \sqrt{117}}{2}$

16. $x^2 - 6x - 5 = 0$

$x^2 - 6x = 5$

$x^2 - 6x + \left(-\frac{3}{2}\right)^2 = 5 + \left(-\frac{3}{2}\right)^2$

$(x - \frac{3}{2})^2 = \frac{20}{4} + \frac{9}{4}$

$x - \frac{3}{2} = \pm\sqrt{\frac{29}{4}}$

$x = \frac{3 \pm \sqrt{29}}{2}$

18. $x^2 - 6x = 3$

$x^2 - 6x + (-3)^2 = 3 + (-3)^2$

$(x-3)^2 = 12$

$x - 3 = \pm\sqrt{12}$

$x = 3 \pm 2\sqrt{3}$

20.
$$x^2 = -x + 5$$
$$x^2 + x = 5$$
$$x^2 + x + \left(\tfrac{1}{2}\right)^2 = 5 + \left(\tfrac{1}{2}\right)^2$$
$$(x + \tfrac{1}{2})^2 = 5 + \tfrac{1}{4}$$
$$(x + \tfrac{1}{2})^2 = \tfrac{21}{4}$$
$$x + \tfrac{1}{2} = \pm\sqrt{\tfrac{21}{4}}$$
$$x = \tfrac{-1 \pm \sqrt{21}}{2}$$

22.
$$2x^2 - 6x + 2 = 0$$
$$x^2 - 3x + 1 = 0$$
$$x^2 - 3x + \left(-\tfrac{3}{2}\right)^2 = -1 + \left(-\tfrac{3}{2}\right)^2$$
$$(x - \tfrac{3}{2})^2 = -\tfrac{4}{4} + \tfrac{9}{4}$$
$$(x - \tfrac{3}{2})^2 = \tfrac{5}{4}$$
$$x - \tfrac{3}{2} = \pm\sqrt{\tfrac{5}{4}}$$
$$x = \tfrac{3 \pm \sqrt{5}}{2}$$

24.
$$2x^2 - 6x + 2 = 0$$
$$x^2 - 3x + 1 = 0$$
$$x^2 - 3x + \left(-\tfrac{3}{2}\right)^2 = -1 + \left(-\tfrac{3}{2}\right)^2$$
$$(x - \tfrac{3}{2})^2 = -\tfrac{4}{4} + \tfrac{9}{4}$$
$$(x - \tfrac{3}{2})^2 = \tfrac{5}{4}$$
$$x - \tfrac{3}{2} = \pm\sqrt{\tfrac{5}{4}}$$
$$x = \tfrac{3 \pm \sqrt{5}}{2}$$

26.
$$2x^2 - 4x + 3 = 0$$
$$x^2 - 2x + \tfrac{3}{2} = 0$$
$$x^2 - 2x + \left(-\tfrac{2}{2}\right)^2 = -\tfrac{3}{2} + \left(-\tfrac{2}{2}\right)^2$$
$$(x - 1)^2 = -\tfrac{3}{2} + \tfrac{2}{2}$$
$$(x - 1)^2 = -\tfrac{1}{2}$$
$$x - 1 = \pm\sqrt{-\tfrac{1}{2}} \times \sqrt{\tfrac{2}{2}}$$
$$x = 1 \pm \tfrac{\sqrt{2}}{2}\, i$$

28.
$$4x^2 - 8x - 1 = 0$$
$$x^2 - 2x - \tfrac{1}{4} = 0$$
$$x^2 - 2x + \left(\tfrac{2}{2}\right)^2 = \tfrac{1}{4} + \left(\tfrac{2}{2}\right)^2$$
$$(x - 1)^2 = \tfrac{1}{4} + 1$$
$$(x - 1)^2 = \tfrac{5}{4}$$
$$x - 1 = \pm\sqrt{\tfrac{5}{4}}$$
$$x = 1 \pm \tfrac{\sqrt{5}}{2}$$
$$x = \tfrac{2}{2} \pm \tfrac{\sqrt{5}}{2}$$
$$x = \tfrac{2 \pm \sqrt{5}}{2}$$

## Exercise 12.2.3

2.
$$3x^2 - 3x = -2$$
$$3x^2 - 3x + 2 = 0$$
$$x = \frac{-(-3) \pm \sqrt{(-3)^2 - 4(3)(2)}}{2(3)}$$
$$x = \frac{3 \pm \sqrt{15i}}{6}$$

4.
$$3x^2 = 3x - 6$$
$$3x^2 - 3x + 6 = 0$$
$$x = \frac{-(-3) \pm \sqrt{(-3)^2 - 4(3)(6)}}{2(3)}$$
$$x = \frac{3 \pm \sqrt{-63}}{6}$$
$$x = \frac{3 \pm \sqrt{-7 * 9}}{6}$$
$$x = \frac{3 \pm 3\sqrt{-7}}{6}$$
$$x = \frac{1 \pm \sqrt{7i}}{2}$$

6. $x^2 + 3x + 1 = 0$

$x = \frac{-(3)\pm\sqrt{(3)^2-4(1)(1)}}{2(1)}$

$x = \frac{-3\pm\sqrt{5}}{2}$

8. $x^2 - 15 = -4x$

$x^2 + 4x - 15 = 0$

$x = \frac{-(4)\pm\sqrt{(4)^2-4(1)(-15)}}{2(1)}$

$x = \frac{-4\pm\sqrt{4*19}}{2}$

$x = \frac{-4\pm2\sqrt{19}}{2}$

$x = -2\pm\sqrt{19}$

10. $x^2 + 4 = 8x$

$x^2 - 8x + 4 = 0$

$x = \frac{-(-8)\pm\sqrt{(-8)^2-4(1)(4)}}{2(1)}$

$x = \frac{8\pm\sqrt{64-16}}{2}$

$x = \frac{8\pm\sqrt{16*3}}{2}$

$x = \frac{8\pm4\sqrt{3}}{2}$

$x = 4\pm2\sqrt{3}$

12. $1 + \frac{7}{x^2} = \frac{3}{x}$

$x^2 + 7 = 3x$

$x^2 - 3x + 7 = 0$

$x = \frac{-(-3)\pm\sqrt{(-3)^2-4(1)(7)}}{2(1)}$

$x = \frac{3\pm\sqrt{9-28}}{2}$

$x = \frac{3\pm\sqrt{-19}}{2}$

$x = \frac{3\pm\sqrt{19}i}{2}$

14. $3x + x(x-3) = 7$

$3x + x^2 - 3x = 7$

$3x + x^2 - 3x - 7 = 0$

$x^2 - 7 = 0$

$x = \frac{-(0)\pm\sqrt{(0)^2-4(1)(-7)}}{2(1)}$

$x = \frac{\pm\sqrt{28}}{2}$

$x = \frac{\pm\sqrt{4*7}}{2}$

$x = \frac{\pm2\sqrt{7}}{2}$

$x = \pm\sqrt{7}$

16. $17x^2 + 5x = 0$

$x = \frac{-(5)\pm\sqrt{(5)^2-4(17)(0)}}{2(17)}$

$x = \frac{-5\pm\sqrt{25}}{34}$

$x = \frac{-5\pm5}{34}$

$x = -\frac{5}{17}$ *or* $x = 0$

18. $36x^2 + 16x - 36 = 0$

$9x^2 + 4x - 9 = 0$

$x = \frac{-(4)\pm\sqrt{(4)^2-4(9)(-9)}}{2(9)}$

$x = \frac{-4\pm\sqrt{16+324}}{18}$

$x = \frac{-4 \pm \sqrt{4*85}}{18}$

$x = \frac{-4 \pm 2\sqrt{85}}{18}$

$x = \frac{-2 \pm \sqrt{85}}{9}$

20. $2x(x+2) + 6x(x-1) = 8$
$2x^2 + 4x + 6x^2 - 6x = 8$
$8x^2 - 2x - 8 = 0$
$4x^2 - x - 4 = 0$

$x = \frac{-(-1) \pm \sqrt{(-1)^2 - 4(4)(-4)}}{2(4)}$

$x = \frac{1 \pm \sqrt{65}}{8}$

22.

$3x(x+3) - (x-1) = (x+5)(x-2)$
$x^2 + x - 20 = 0$

$x = \frac{-(1) \pm \sqrt{(1)^2 - 4(1)(-20)}}{2(1)}$

$x = \frac{-1 \pm \sqrt{81}}{2}$

$x = \frac{-1 \pm 9}{2}$

$x = 4$ *or* $x = -5$

24. $5x^2 = 8x - 7$
$5x^2 - 8x + 7 = 0$

$x = \frac{-(-8) \pm \sqrt{(-8)^2 - 4(5)(7)}}{2(5)}$

$x = \frac{8 \pm \sqrt{4*19}}{10}$

$x = \frac{4 \pm \sqrt{19}}{5}$

26. $x^2 + 8 = 3x$
$x^2 - 3x + 8 = 0$

$x = \frac{-(-3) \pm \sqrt{(-3)^2 - 4(1)(8)}}{2(1)}$

$x = \frac{3 \pm \sqrt{-23}}{2}$

$x = \frac{3 \pm \sqrt{23}i}{2}$

28. $4y^2 - (y-1)(y+3) = 10$
$-y^2 + 2y - 7 = 0$

$x = \frac{-(2) \pm \sqrt{(2)^2 - 4(-1)(-7)}}{2(-1)}$

$x = 1 \pm \sqrt{6}\,i$

Exercise 12.2.4

2. $A = P(1+r)^t$
$A = 50000(1 + 0.085)^{21}$

$A \doteq 277300$

The amount at the end of 21 years will be \$277300 (approximately).

4. $d = 4.9t^2$
$100 = 4.9t^2$
$\pm\sqrt{20.408} = t$
$4.52 = t$

The book will fall for approximately 4.52 seconds prior to hitting the ground.

6. $A = P(1+r)^t$
$A = 250000(1.102)^{15}$
$A \doteq 1073150$

The amount at the end of 15 years will be 1073150 (approximately).

Exercise 12.3.1

2. $5x^2 + 11x + 21 = 0$

$$\begin{aligned} &b^2 - 4ac \\ &= (11)^2 - 4(5)(21) \\ &= -299 \end{aligned}$$

The discriminant is negative, therefore the solutions are two non-real complex numbers.

4. $x^2 - x + 1 = 0$

$$\begin{aligned} &b^2 - 4ac \\ &= (-1)^2 - 4(1)(1) \\ &= -3 \end{aligned}$$

The discriminant is negative, therefore the solutions are two non-real complex numbers.

6. $x^2 - 3 = 0$

$$\begin{aligned} &b^2 - 4ac \\ &= (0)^2 - 4(1)(-3) \\ &= 12 \end{aligned}$$

The discriminant is positive, therefore the solutions are two different real-numbers.

8. $2x^2 - 3x - 3 = 0$

$$\begin{aligned} &b^2 - 4ac \\ &= (-3)^2 - 4(2)(-3) \\ &= 33 \end{aligned}$$

The discriminant is positive, therefore the solutions are two different real-numbers.

10. $x^2 + 4x + 2 = 0$

$$\begin{aligned} &b^2 - 4ac \\ &= (4)^2 - 4(1)(2) \\ &= 8 \end{aligned}$$

The discriminant is positive, therefore the solutions are two different real-numbers.

12. $12t^2 - 9t = 0$

$$\begin{aligned} &b^2 - 4ac \\ &= (-9)^2 - 4(12)(0) \\ &= 81 \end{aligned}$$

The discriminant is positive, therefore the solutions are two different real-numbers.

14. $x^2 + \frac{3}{5}x + 7 = 0$

$5x^2 + 3x + 35 = 0$

$$\begin{aligned} &b^2 - 4ac \\ &= (3)^2 - 4(5)(25) \\ &= -691 \end{aligned}$$

The discriminant is negative, therefore the solutions are two non-real complex numbers.

16. $7y^2 - 3\sqrt{3}y - 5 = 0$

$$\begin{aligned} &b^2 - 4ac \\ &= \left(-3\sqrt{3}\right)^2 - 4(7)(-5) \\ &= 167 \end{aligned}$$

The discriminant is positive, therefore the solutions are two different real-numbers.

Exercise 12.3.2

2. $x = 3 \;\; and \quad x = 7$

$x - 3 = 0 \qquad x - 7 = 0$

$(x - 3)(x - 7) = 0$

$x^2 - 10x + 21 = 0$

4. $x = 4$ *and* $x = 6$

$x - 4 = 0 \quad x - 6 = 0$

$(x-4)(x-6)=0$

$x^2 - 10x + 24 = 0$

6. $x = 9i$ *and* $x = -9i$

$x - 9i = 0 \quad x + 9i = 0$

$(x-9i)(x+9i)=0$

$x^2 + 81 = 0$

8. $x = 7\sqrt{5}$ *and* $x = \sqrt{5}$

$x - 7\sqrt{5} = 0 \quad x - \sqrt{5} = 0$

$\left(x - 7\sqrt{5}\right)\left(x - \sqrt{5}\right)=0$

$x^2 - 8\sqrt{5}x + 35 = 0$

Exercise 12.4.1

2. $(2x)\left[\frac{x}{2} - \frac{4}{x}\right] = [1]\ (2x)$

$x^2 - 8 = 2x$

$x^2 - 2x - 8 = 0$

$(x-4)(x+2) = 0$

$x = 4$ *or* $x = -2$

Check the solutions.

$x = 4$

| $LHS$ | $RHS$ |
|---|---|
| $\frac{x}{2} - \frac{4}{x}$ | 1 |
| $= \frac{4}{2} - \frac{4}{4}$ | |
| $= 2 - 1$ | |
| $= 1$ | |

LHS = RHS

$x = -2$

| $LHS$ | $RHS$ |
|---|---|
| $\frac{x}{2} - \frac{4}{x}$ | 1 |
| $= \frac{-2}{2} - \frac{4}{-2}$ | |
| $= -1 - (-2)$ | |
| $= -1 + 2$ | |
| $= 1$ | |

LHS = RHS

The solutions check.

4.

$(2)(8\text{-}x)(4\text{+}x)\left[\frac{1}{8-x} - \frac{1}{4+x}\right]$

$= \left[\frac{1}{2}\right](8\text{-}x)(2)(4\text{+}x)$

$(2)(4+x) - (2)(8-x)$

$= (8-x)(4+x)$

$0 = -x^2 + 40$

$x = \frac{-(0)\pm\sqrt{(0)^2-4(-1)(40)}}{2(-1)}$

$x = \pm 2\sqrt{10}$

The solutions check.

6.

$(4x)\left[\frac{x+3}{x}\right] = \left[\frac{x-2}{4}\right](4x)$

$4(x+3) = x(x-2)$

$4x + 12 = x^2 - 2x$

$0 = x^2 - 6x - 12$

$x = \frac{-(-6)\pm\sqrt{(-6)^2-4(1)(-12)}}{2(1)}$

$x = 3 \pm \sqrt{21}$

The solutions check.

8.

$$(6)(8)(x)\left[\frac{x}{6}-\frac{6}{x}\right]=\left[\frac{x}{8}\right](6)(8)(x)$$

$$8x^2-288=6x^2$$
$$x^2=144$$
$$x=\pm 12$$

$x=12$ *or* $x=-12$

The solutions check.

10.

$$(x+3)(x-6)\left[\frac{2x}{x+3}-\frac{3}{x-6}\right]$$

$$=\left[\frac{29}{(x-6)(x+3)}\right](x+3)(x-6)$$

LCM $=(x+3)(x-6)$

$$2x(x-6)-3(x+3)=29$$

$$2x^2-15x-38=0$$

Substitute the values of : $a=2$, $b=-15$, $c=-38$ into the quadratic formula.

$$x=\frac{-(-15)\pm\sqrt{(-15)^2-4(2)(-38)}}{2(2)}$$

$x=9.5$ *or* $x=-2$

The solutions check.

12.

$$(2)\left[\frac{x+1}{2}+\frac{x+3}{2}\right]=\left[\frac{x+4}{2}\right](2)$$

$$x+1+x+3=x+4$$

$$x=0$$

The solution checks.

14. $$(4)\left[2x-\frac{x^2}{4}\right]=\left[6x-9\right](4)$$

$$8x-x^2=24x-36$$

$$0=x^2+16x-36$$

$$0=(x+18)(x-2)$$

$x=-18$ *or* $x=2$

The solutions check.

Exercise 12.4.2

2.

| | Case 1 | Case2 |
|---|---|---|
| Dist.(km) | 500 | 500 |
| Speed($\frac{\text{km}}{\text{hr}}$) | $x$ | $x+5$ |
| time (hr) | $t$ | $t-5$ |

$$v=\frac{d}{t}$$

$$x=\frac{500}{t} \quad EQ1$$

$$x+5=\frac{500}{t-5} \quad EQ2$$

Substitute EQ1 into EQ2.

$$(t)(t-5)\left[\frac{500}{t}+5\right]=\left[\frac{500}{t-5}\right](t)(t-5)$$

$$500(t-5)+5t^2-25t=500t$$

$$500t-2500+5t^2-25t-500t=0$$

$$5t^2-25t-2500=0$$

$$t^2-5t-500=0$$
$$(t+20)(t-25)=0$$

$t=-20$ *or* $t=25$

Since time cannot be negative, t=25, and checks. Substituting this value into EQ1, we obtain the speed of the cyclist.

$x = \frac{500}{t}$ $EQ1$

$x = \frac{500}{25}$

The speed of the cyclist is 20 $\frac{\text{km}}{\text{hr}}$.

4.

| | Case 1 | Case2 |
|---|---|---|
| Dist.(km) | 100 | 100 |
| Speed($\frac{\text{km}}{\text{hr}}$) | $x$ | $x-5$ |
| time (hr) | $t$ | $t+1$ |

$v = \frac{d}{t}$

$x = \frac{100}{t}$ $EQ1$

$x - 5 = \frac{100}{t+1}$ $EQ2$

Substitute EQ1 into EQ2.

$(t)(t+1)\left[\frac{100}{t} - 5\right] = \left[\frac{100}{t+1}\right](t)(t+1)$

$100(t+1) - 5(t)(t+1) = 100t$

$100t + 100 - 5t^2 - 5t = 100t$

$-5t^2 - 5t + 100 = 0$

divide by: $-5$ on both sides

$t^2 + t - 20 = 0$
$(t+5)(t-4) = 0$

$t = -5$ $or$ $t = 4$

Since time cannot be negative, t=4, and checks. Substituting this value into EQ1, we obtain the speed of the cyclist.

$x = \frac{100}{t}$ $EQ1$

$x = \frac{100}{4}$

The speed of the cyclist is 25 $\frac{\text{km}}{\text{hr}}$.

6.

| | Case 1 | Case2 |
|---|---|---|
| Dist.(km) | 1000 | 1500 |
| Speed($\frac{\text{km}}{\text{hr}}$) | $s$ | $s-250$ |
| time (hr) | $t$ | $t+1$ |

$s = \frac{d}{t} \Rightarrow t = \frac{d}{s}$

$s = \frac{1000}{t}$ $EQ1$

$s - 250 = \frac{1500}{t+1}$ $EQ2$

Substitute EQ1 into EQ2.

$(t)(t+1)\left[\frac{1000}{t} - 250\right] = \left[\frac{1500}{t+1}\right](t)(t+1)$

$1000(t+1) - 250(t)(t+1) = 1500t$

$1000t + 1000 - 250t^2 - 250t = 1500t$

$1000t + 1000 - 250t^2 - 250t - 1500t = 0$

$-250t^2 - 750t + 1000 = 0$

divide both sides by: $-250$

$t^2 + 3t - 4 = 0$
$(t+4)(t-1) = 0$

$t = -4$ $or$ $t = 1$

Since the time cannot be negative, t=1, and checks. Substituting this value into the above equations for speed, we obtain the speed of each plane. The first plane at 1000$\frac{\text{km}}{\text{hr}}$ and the second plane at 750 $\frac{\text{km}}{\text{hr}}$.

8.

| | Upstream | Downstream |
|---|---|---|
| Dist.(km) | 100 | 100 |
| Speed($\frac{\text{km}}{\text{hr}}$) | $s$ - 5 | $s$+5 |
| time (hr) | $t_1$ | $t_2$ |

$$s = \frac{d}{t} \Rightarrow t = \frac{d}{s}$$

$$t_1 + t_2 = 10\tfrac{2}{3}$$

$$t_1 = \frac{100}{s-5} \qquad EQ1$$

$$t_2 = \frac{100}{s+5} \qquad EQ2$$

Substitute all values into: $t_1 + t_2 = 10\frac{2}{3}$

$$(s-5)(s+5)\left[\frac{100}{s-5} + \frac{100}{s+5}\right] = \left[10\tfrac{2}{3}\right](s-5)(s+5)$$

$$0 = 32s^2 - 600s - 800$$
$$0 = 4s^2 - 75s - 100$$

$s = 20 \qquad or \qquad s = -1.25$

Since the speed cannot be negative, s = 20, and checks. The speed of the boat in still water is 20 $\frac{\text{km}}{\text{hr}}$.

10.
$\frac{t}{a} + \frac{t}{b} = 1$

The time taken by Ted and Ling is 8 hrs. The time taken by Ted alone is $(x+4)$ hrs. The time taken by Ling alone is $x$ hrs.

$$(x)(x+4)\left[\frac{8}{x} + \frac{8}{x+4}\right] = \left[1\right](x)(x+4)$$

$$0 = x^2 - 12x - 32$$

$x = 14.25 \qquad or \qquad x = -2.25$

Since time cannot be negative, $x = 14.25$, which checks.
It would take Ling 14.25 hours and Ted 18.25 hours to tile the floor individually.

Exercise 12.5.1

2.

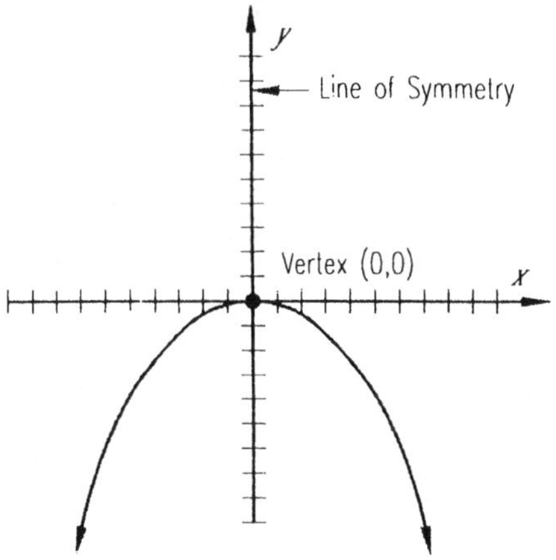


$y = -\frac{1}{6}x^2$

4.

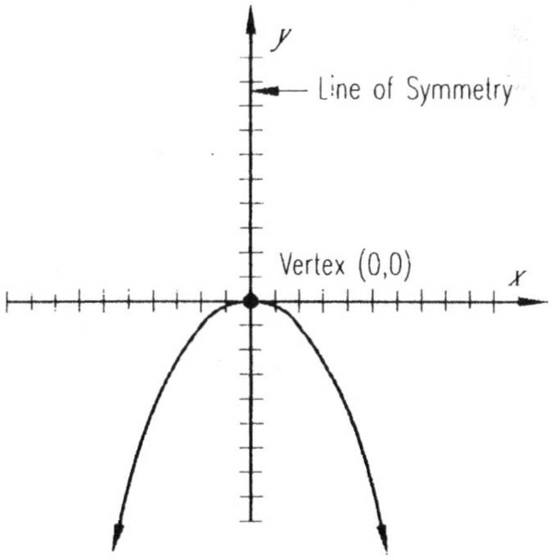


$y = -\frac{1}{3}x^2$

6.

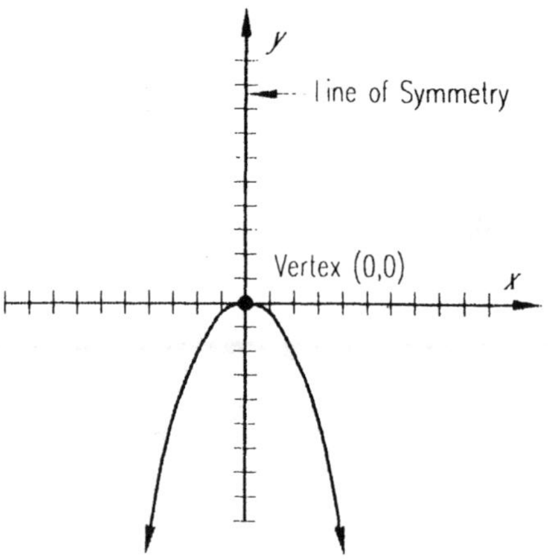


$y = -\frac{1}{2}x^2$

8.

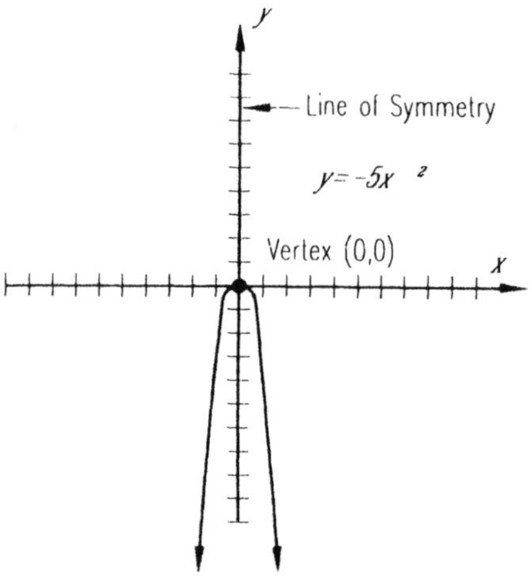


## Exercise 12.5.2

2.

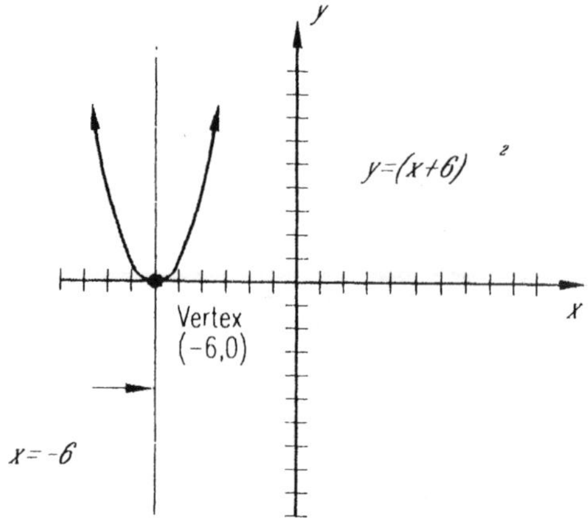


4.

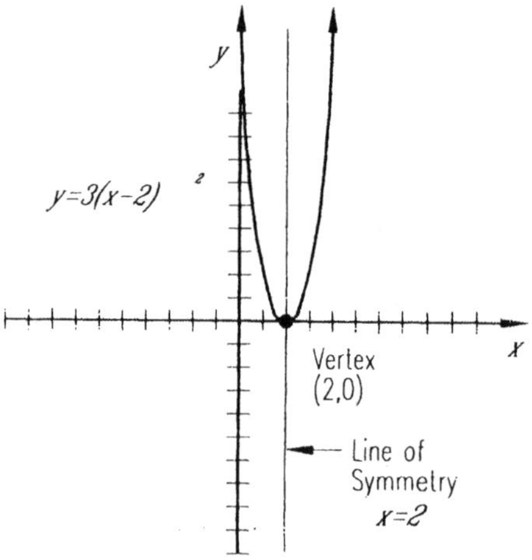

# CHAPTER 13: Algebraic Fractions

## Exercise 13.1.1

2. $x^2 + 5x - 6 = 0$
$(x+6)(x-1) = 0$

$(x+6) = 0$ *or* $(x-1) = 0$
$x = -6$ $x = 1$

4. $x^2 - x - 6 = 0$
$(x-3)(x+2) = 0$

$(x-3) = 0$ *or* $(x+2) = 0$
$x = 3$ $x = -2$

6. $x^2 - 9x + 20 = 0$
$(x-5)(x-4) = 0$

$(x-5) = 0$ *or* $(x-4) = 0$
$x = 5$ $x = 4$

## Exercise 13.1.2

2. $\frac{2x^3}{14x^2} = \frac{2x^2 x}{2*7x^2} = \frac{x}{7}$

4. $\frac{78y^3}{13y^6} = \frac{13*6y^3}{13y^3y^3} = \frac{6}{y^3}$

6. $\frac{9x+27}{3} = \frac{3(3x+9)}{3} = 3x+9$
$= 3(x+3)$

8. $\frac{2t-24}{12} = \frac{2(t-12)}{2*6} = \frac{t-12}{6}$

10. $\frac{t-2}{-t+2} = \frac{t-2}{-(t-2)} = \frac{1}{-1}*\frac{t-2}{t-2} = -1$

12. $\frac{x^2-x-2}{x^2+4x-12} = \frac{(x-2)(x+1)}{(x+6)(x-2)} = \frac{x+1}{x+6}$

14. $\frac{t^2-9}{5t^2-14t-3} = \frac{(t+3)(t-3)}{(5t+1)(t-3)} = \frac{t+3}{5t+1}$

16. $\frac{x^2-x-12}{x^2-9} = \frac{(x-4)(x+3)}{(x-3)(x+3)} = \frac{x-4}{x-3}$

18.
$\frac{a^3+b^3}{a+b} = \frac{(a+b)(a^2-ab+b^2)}{a+b} = a^2 - ab + b^2$

## Exercise 13.2.1

2. $\frac{5x^2}{15x+1}*\frac{45x+5}{5x^2} = \frac{(5x^2)5(15x+1)}{(15x+1)(5x^2)} = 5$

4. $\frac{x^2-x-12}{x^2-9}*\frac{x^2+3x-18}{x^2-2x-8} = \frac{(x-4)(x+3)(x-3)(x+6)}{(x-3)(x+3)(x-4)(x+2)}$

$= \frac{x+6}{x+2}$

6. $\frac{x^2-4x}{x^2-16}*\frac{x^2+7x+12}{x+3} = \frac{x(x-4)(x+4)(x+3)}{(x+4)(x-4)(x+3)} = x$

8. $\frac{t^2-4}{2t+4}*\frac{t-3}{t-2} = \frac{(t-2)(t+2)(t-3)}{2(t+2)(t-2)} = \frac{t-3}{2}$

10. $\frac{x^2-x-42}{x^2-x-6}*\frac{x^2-2x-3}{x^2-4x-21} = \frac{(x+6)(x-7)(x-3)(x+1)}{(x-3)(x+2)(x-7)(x+3)}$

$= \frac{(x+6)(x+1)}{(x+2)(x+3)}$

12. $\frac{a^2-b^2}{3a+3b}*\frac{a+b}{a-b} = \frac{(a-b)(a+b)(a+b)}{3(a+b)(a-b)}$

$= \frac{a+b}{3}$

## Exercise 13.2.2

2. $\frac{8x^2}{2y^2} \div \frac{16x}{4y^2} = \frac{(8x)(x)2(2y^2)}{(2y^2)2(8x)} = x$

4. $\frac{x^3-27}{x^3+8} \div \frac{x-3}{x^2-2x+4} = \frac{x^3-3^3}{x^3+2^3}*\frac{x^2-2x+4}{x-3}$

$= \frac{(x-3)(x^2+3x+9)(x^2-2x+4)}{(x+2)(x^2-2x+4)(x-3)} = \frac{x^2+3x+9}{x+2}$

6.
$\frac{t^2-25}{t^2+t-2} \div \frac{t^2+2t-35}{t^2+6t-7} = \frac{(t-5)(t+5)(t+7)(t-1)}{(t+2)(t-1)(t+7)(t-5)}$

$= \frac{t+5}{t+2}$

8.
$\frac{9x^2-4}{x^2-16} \div \frac{6x+4}{x+4} = \frac{(3x)^2-(2)^2}{(x-4)(x+4)}*\frac{x+4}{2(3x+2)}$

$= \frac{(3x-2)(3x+2)(x+4)}{(x-4)(x+4)2(3x+2)}$

$= \frac{3x-2}{2(x-4)}$

10.

$$\frac{x^2+5x-14}{x^2+x-12} \div \frac{x^2-x-2}{x^2+5x+4}$$

$$= \frac{(x+7)(x-2)(x+4)(x+1)}{(x+4)(x-3)(x-2)(x+1)}$$

$$= \frac{x+7}{x-3}$$

12. $\frac{2y^3}{4x^2} \div \frac{2y^2}{2x^3} = \frac{(2y^2)(y)(2x^2)(x)}{2(2x^2)(2y^2)} = \frac{xy}{2}$

## Exercise 13.2.3

2.

$$x^2 - x - 6 = (x-3)(x+2)$$

$$x^2 + 3x + 2 = (x+1)(x+2)$$

$$LCM = (x-3)(x+2)(x+1)$$

4.

$$6v^2t = 2v(3vt)$$

$$3vt^2 = t(3vt)$$

$$LCM = 2v(3vt)(t) = 6v^2t^2$$

6.

$$x^2 - 9 = (x-3)(x+3)$$

$$3x - 9 = 3(x-3)$$

$$LCM = 3(x-3)(x+3)$$

8.

$$x^2 - 2x - 8 = (x-4)(x+2)$$

$$x^2 + 4x + 3 = (x+3)(x+1)$$

$$LCM = (x-4)(x+2)(x+3)(x+1)$$

10.

$$5x^2 - 5x = 5x(x-1)$$

$$5x^2 = (5x)x$$

$$x^2 - 1 = (x-1)(x+1)$$

$$LCM = (x-1)(x+1)(5x)x$$
$$= (5x^2)(x-1)(x+1)$$

12.

$$x + 2 = x + 2$$

$$x^2 + x - 6 = (x+3)(x-2)$$

$$x^2 - 4 = (x+2)(x-2)$$

$$LCM = (x+3)(x-2)(x+2)$$

14.

$$x^2 - x - 20 = (x-5)(x+4)$$

$$x^2 + x - 12 = (x+4)(x-3)$$

$$LCM = (x+4)(x-3)(x-5)$$

## Exercise 13.2.4

2.

$$\frac{2x}{x^2-y^2} + \frac{x+4}{x^2-y^2}$$

$$= \frac{2x+x+4}{x^2-y^2}$$

$$= \frac{3x+4}{x^2-y^2}$$

4.

$\frac{x}{x^2-16} + \frac{x+1}{x-4}$

$x^2 - 16 = (x-4)(x+4)$

$x - 4 = x - 4$

$LCM = (x-4)(x+4)$

$= \frac{x+x^2+5x+4}{(x-4)(x+4)}$

$= \frac{x^2+6x+4}{(x-4)(x+4)}$

$= \frac{x^2+6x+4}{x^2-16}$

6.

$\frac{4y}{2y^2+4y+2} + \frac{9}{3y^2+5y+2}$

$2y^2 + 4y + 2 = (2y+2)(y+1)$

$3y^2 + 5y + 2 = (y+1)(3y+2)$

$LCM = (2y+2)(y+1)(3y+2)$

$= \frac{4y}{(2y+2)(y+1)} \times \frac{3y+2}{3y+2} + \frac{9}{(y+1)(3y+2)} \times \frac{2y+2}{2y+2}$

$= \frac{12y^2+8y+18y+18}{(2y+2)(y+1)(3y+2)}$

$= \frac{12y^2+26y+18}{(2y+2)(y+1)(3y+2)}$

$= \frac{2(6y^2+13y+9)}{(2y+2)(y+1)(3y+2)}$

$= \frac{2(6y^2+13y+9)}{2(y+1)(y+1)(3y+2)}$

$= \frac{6y^2+13y+9}{(y+1)(y+1)(3y+2)}$

8.

$\frac{1}{x^2-25} + \frac{5x}{x+5} + \frac{5}{x-5}$

$x^2 - 25 = (x+5)(x-5)$

$x + 5 = x + 5$

$x - 5 = x - 5$

$LCM = (x+5)(x-5)$

$= \frac{1}{(x-5)(x+5)} + \frac{5x}{x+5} \times \frac{x-5}{x-5} + \frac{5}{x-5} \times \frac{x+5}{x+5}$

$= \frac{1+5x^2-25x+5x+25}{(x+5)(x-5)}$

$= \frac{5x^2-20x+26}{(x+5)(x-5)}$

Exercise 13.2.5

2.

$\frac{7y}{y} - \frac{8}{y} = \frac{7y-8}{y}$

4.

$\frac{y^2}{3xy^2} - \frac{xy}{3xy^2}$

$= \frac{y^2-xy}{3xy^2}$

$= \frac{y(y-x)}{3xy^2}$

$= \frac{y-x}{3xy}$

6.

$\frac{6x}{x^2-4} + \frac{2}{2-x} - \frac{2}{x+2}$

$x^2 - 4 = (x+2)(x-2)$

$2 - x = -1(x-2)$

$x + 2 = x + 2$

$LCM = (x+2)(x-2)$

$= \frac{6x}{(x+2)(x-2)} + \frac{2}{-1(x-2)} \times \frac{x+2}{x+2} - \frac{2}{x+2} \times \frac{x-2}{x-2}$

$= \frac{6x-2(x+2)-2(x-2)}{(x+2)(x-2)}$

$= \frac{6x-2x-4-2x+4}{(x+2)(x-2)}$

$= \frac{2x}{(x+2)(x-2)}$

$= \frac{2x}{x^2-4}$

8.

$\frac{2x}{x^2+3x-18} - \frac{x+6}{x^2-2x-3}$

$x^2 + 3x - 18 = (x-3)(x+6)$

$x^2 - 2x - 3 = (x-3)(x+1)$

$LCM = (x+1)(x-3)(x+6)$

$= \frac{2x}{(x-3)(x+6)} \times \frac{x+1}{x+1} - \frac{x+6}{(x-3)(x+1)} \times \frac{x+6}{x+6}$

$= \frac{2x^2+2x-(x^2+12x+36)}{(x+6)(x-3)(x+1)}$

$= \frac{2x^2+2x-x^2-12x-36}{(x+6)(x-3)(x+1)}$

$= \frac{x^2-10x-36}{(x+6)(x-3)(x+1)}$

## Exercise 13.3.1

2.

$\frac{2}{3} + \frac{1}{4} = \frac{x}{12}$

$LCM = (3)(4) = 12$

$(12) \quad [\frac{2}{3} + \frac{1}{4}] = [\frac{x}{12}](12)$

$\frac{2}{3}(12) + \frac{1}{4}(12) = \frac{x}{12}(12)$

$8 + 3 = x$

$11 = x$

| *LHS* | *RHS* |
|---|---|
| $\frac{2}{3} + \frac{1}{4}$ | $\frac{x}{12}$ |
| $= \frac{8+3}{12}$ | $= \frac{11}{12}$ |
| $= \frac{11}{12}$ | |

The LHS=RHS therefore the solution is $x = 11$.

4.

$\frac{1}{9} - \frac{2}{3} = \frac{1}{x}$

$LCM = 9x$

$(9x) \quad [\frac{1}{9} - \frac{2}{3}] = [\frac{1}{x}](9x)$

$\frac{1}{9}(9x) - \frac{2}{3}(9x) = \frac{1}{x}(9x)$

$x - 6x = 9$

$-5x = 9$

$x = -\frac{9}{5}$

| *LHS* | *RHS* |
| --- | --- |
| $\frac{1}{9} - \frac{2}{3}$ | $\frac{1}{x}$ |
| $= \frac{1-6}{9}$ | $= \frac{1}{-\frac{9}{5}}$ |
| $= -\frac{5}{9}$ | $= -\frac{5}{9}$ |

The LHS=RHS therefore the solution is $x = -\frac{9}{5}$.

6. $\frac{x}{8} - \frac{x}{5} = 7$

$LCM = 40$

$$(40) \left[\frac{x}{8} - \frac{x}{5}\right] = [7](40)$$

$$\frac{x}{8}(40) - \frac{x}{5}(40) = 7\,(40)$$

$$5x - 8x = 280$$

$$-3x = 280$$

$$x = -\frac{280}{3}$$

The LHS=RHS therefore the solution is $x = -\frac{280}{3}$.

8. $\frac{9}{t} + 4 = -6$

$LCM = t$

$$(t)\left[\frac{9}{t} + 4\right] = [-6](t)$$

$$\frac{9}{t}(t) + 4(t) = -6\,(t)$$

$$9 + 4t = -6t$$

$$-\frac{9}{10} = t$$

The LHS=RHS therefore the solution is $t = -\frac{9}{10}$.

10.

$$\frac{9}{3r} + \frac{1}{r} = \frac{12}{5}$$

$LCM = 15r$

$$(15r)\left[\frac{9}{3r} + \frac{1}{r}\right] = \left[\frac{12}{5}\right](15r)$$

$$\frac{9}{3r}(15r) + \frac{1}{r}(15r) = \frac{12}{5}\,(15r)$$

$$45 + 15 = 36r$$

$$60 = 36r$$

$$\frac{5}{3} = r$$

The LHS=RHS therefore the solution is $r = \frac{5}{3}$.

12.

$$\frac{x+3}{5+x} = \frac{7}{8}$$

$LCM = (5+x)(8)$

$$(8)(5+x)\left[\frac{x+3}{5+x}\right] = \left[\frac{7}{8}\right](8)(5+x)$$

$$8(x+3) = 7(5+x)$$

$$8x + 24 = 35 + 7x$$

$$x = 11$$

The LHS=RHS therefore the solution is $x = 11$.

14.

$$\frac{3}{x} - \frac{5}{x} + \frac{8}{x} = 8$$

$LCM = x$

$$(x)\left[\frac{3}{x} - \frac{5}{x} + \frac{8}{x}\right] = [8](x)$$

$$3 - 5 + 8 = 8x$$

$$6 = 9x$$

$$x = \frac{2}{3}$$

The LHS=RHS therefore the solution is $x = \frac{2}{3}$.

16.

$$\frac{1}{3} - \frac{1}{3x} = \frac{8}{x} + \frac{3}{5}$$

$$LCM = 15x$$

$$(15x)\left[\frac{1}{3} - \frac{1}{3x}\right] = \left[\frac{8}{x} + \frac{3}{5}\right](15x)$$

$$5x - 5 = 120 + 9x$$

$$-\frac{125}{4} = x$$

$$x = -31.25$$

The LHS=RHS therefore the solution is $x = -31.25$.

18.

$$\frac{7}{x} - \frac{7}{x-3} = \frac{5}{x}$$

$$LCM = x(x-3)$$

$$(x)(x-3)\left[\frac{7}{x} - \frac{7}{x-3}\right] = \left[\frac{5}{x}\right](x)(x-3)$$

$$\left(\frac{7}{x}\right)(x)(x-3) - \left(\frac{7}{x-3}\right)(x)(x-3) = \frac{5}{x}(x)(x-3)$$

$$7x - 21 - 7x = 5x - 15$$

$$-21 = 5x - 15$$

$$x = -\frac{6}{5}$$

The LHS=RHS therefore the solution is $x = -\frac{6}{5}$.

20.

$$\frac{7}{3x+3} = \frac{4}{x+5}$$

$$LCM = (3x+3)(x+5)$$

$$(3x+3)(x+5)\left[\frac{7}{3x+3}\right] = \left[\frac{4}{x+5}\right](3x+3)(x+5)$$

$$7(x+5) = 4(3x+3)$$

$$7x + 35 = 12x + 12$$

$$23 = 5x$$

$$x = \frac{23}{5}$$

The LHS=RHS therefore the solution is $x = \frac{23}{5}$.

22.

$$\frac{5}{x-2} + 4 = \frac{20}{x^2-4} + \frac{25}{x+2}$$

$$LCM = (x-2)(x+2)$$

$$(x-2)(x+2)\left[\frac{5}{x-2} + 4\right]$$

$$= \left[\frac{20}{x^2-4} + \frac{25}{x+2}\right](x-2)(x+2)$$

$$5(x+2) + 4x^2 - 16 = 20 + 25(x-2)$$

$$5x + 10 + 4x^2 - 16 = 20 + 25x - 50$$

$$4x^2 - 20x + 24 = 0$$

$$x^2 - 5x + 6 = 0$$

$$(x-3)(x-2) = 0$$

thus:

$x - 3 = 0$ or $x - 2 = 0$

$x = 3$ $\quad$ $x = 2$

Substituting the values of $x = 3$ and $x = 2$ into the original equation shows that $x = 2$ **is not a solution** since it produces division by zero.
The LHS=RHS therefore the solution is $x = 3$.

24.

$$\frac{7x-1}{x^2+3x+2} = \frac{4x-7}{x+2}$$

Note: $x^2 + 3x + 2 = (x+2)(x+1)$

$LCM = (x+2)(x+1)$

$$(x+2)(x+1)\left[\frac{7x-1}{x^2+3x+2}\right] = \left[\frac{4x-7}{x+2}\right](x+2)(x+1)$$

$$7x - 1 = (4x-7)(x+1)$$
$$7x - 1 = 4x^2 + 4x - 7x - 7$$

$$0 = 4x^2 - 10x - 6$$

$$0 = (2x-6)(2x+1)$$

thus:

$2x - 6 = 0$ or $2x + 1 = 0$

$x = 3$ $\quad x = -\frac{1}{2}$

The LHS=RHS for both values of $x$, therefore the solutions are $x = 3$ and $x = -\frac{1}{2}$.

26.

$$\frac{6}{x-1} + 2 = \frac{16}{x^2-1} + \frac{12}{x+1}$$

$LCM = (x-1)(x+1)$

$$(x-1)(x+1)\left[\frac{6}{x-1} + 2\right]$$

$$= \left[\frac{16}{x^2-1} + \frac{12}{x+1}\right](x-1)(x+1)$$

$$6(x+1) + 2(x^2-1) = 16 + 12(x-1)$$

$$6x + 6 + 2x^2 - 2 = 16 + 12x - 12$$

$$2x^2 - 6x = 0$$

$$x^2 - 3x = 0$$

$$x(x-3) = 0$$

thus:

$x = 0$ or $x - 3 = 0$

$x = 0$ $\quad x = 3$

The LHS=RHS for both values of $x$, therefore the solutions are $x = 0$ and $x = 3$.

28.

$$\frac{6x}{x+1} + \frac{24x}{x^2+7x+6} = 6$$

Note: $x^2 + 7x + 6 = (x+6)(x+1)$

$LCM = (x+6)(x+1)$

$$(x+6)(x+1)\left[\frac{6x}{x+1} + \frac{24x}{x^2+7x+6}\right] = [6](x+6)(x+1)$$

$$6x(x+6) + 24x = 6(x^2 + 7x + 6)$$

$$6x^2 + 36x + 24x = 6x^2 + 42x + 36$$

$$18x = 36$$

$$x = 2$$

The LHS=RHS therefore the solution is $x = 2$.

Exercise 13.3.2

2.

Evaluate

'a number' $\Rightarrow x$

'twenty times it's reciprocal $\Rightarrow 20 \times \frac{1}{x}$

Interpret

$$x + \frac{20}{x} = 12$$

Solve

$$x + \frac{20}{x} = 12$$

$$(x)\left[x + \frac{20}{x}\right] = \left[12\right](x)$$

$$x^2 + 20 = 12x$$

$$x^2 - 12x + 20 = 0$$

$$(x-2)(x-10) = 0$$

thus:

$x - 2 = 0$ or $x - 10 = 0$

$x = 2$ $x = 10$

Check

$x = 2$

| LHS | RHS |
|---|---|
| $x + \frac{20}{x}$ | 12 |
| $= 2 + \frac{20}{2}$ | |
| $= 2 + 10$ | |
| $= 12$ | |

LHS=RHS

$x = 10$

| LHS | RHS |
|---|---|
| $x + \frac{20}{x}$ | 12 |
| $= 10 + \frac{20}{10}$ | |
| $= 10 + 2$ | |
| $= 12$ | |

LHS=RHS

Close

The numbers are 2 and 10.

4.

Evaluate

' reciprocal of 10' $\Rightarrow \frac{1}{10}$

' reciprocal of 7' $\Rightarrow \frac{1}{7}$

'reciprocal of what number' $\Rightarrow \frac{1}{x}$

Interpret

$$\frac{1}{10} + \frac{1}{7} = \frac{1}{x}$$

Solve

$$\frac{1}{10} + \frac{1}{7} = \frac{1}{x}$$

$$(70x)\left[\frac{1}{10} + \frac{1}{7}\right] = \left[\frac{1}{x}\right](70x)$$

$$7x + 10x = 70$$

$$17x = 70$$

$$x = \frac{70}{17}$$

Check

$x = \frac{70}{17}$

| LHS | RHS |
|---|---|
| $\frac{1}{10} + \frac{1}{7}$ | $\frac{1}{x}$ |
| $= \frac{7+10}{70}$ | $= \frac{1}{\frac{70}{17}}$ |
| $= \frac{17}{70}$ | $= \frac{17}{70}$ |

LHS=RHS

Close

The reciprocal of 10 plus the reciprocal of 7 is the reciprocal of $\frac{70}{17}$.

6.

Evaluate

'a number' $\Rightarrow x$
'nine times it's reciprocal $\Rightarrow 9 \times \frac{1}{x}$

Interpret

$$x + \frac{9}{x} = 6$$

Solve

$$x + \frac{9}{x} = 6$$

$$(x)\left[x + \frac{9}{x}\right] = \left[6\right](x)$$

$$x^2 + 9 = 6x$$

$$x^2 - 6x + 9 = 0$$

$$(x-3)(x-3) = 0$$

thus:

$x - 3 = 0$ or $x - 3 = 0$

$x = 3$ $x = 3$

Check

$x = 3$

| LHS | RHS |
|---|---|
| $x + \frac{9}{x}$ | 6 |
| $= 3 + \frac{9}{3}$ | |
| $= 3 + 3$ | |
| $= 6$ | |

LHS=RHS

Close
The number is 3.

8.

Evaluate

| | Car | Truck |
|---|---|---|
| Distance | 180 km | 150 km |
| Speed | $s + 20$ | $s$ |
| time | $t$ | $t$ |

Interpret

$$\text{speed} = \frac{\text{distance}}{\text{time}}$$

$$s = \frac{d}{t}$$

$EQ1(\text{car})$ $s + 20 = \frac{180}{t}$

$EQ2(\text{truck})$ $s = \frac{150}{t}$

Solve

Substitute EQ2 into EQ1.

EQ1 $s + 20 = \frac{180}{t}$

$(t)\left[\left(\frac{150}{t}\right) + 20\right] = \left[\frac{180}{t}\right](t)$

$150 + 20t = 180$

$20t = 30$

$t = \frac{3}{2}$

$t = 1.5$

Speed of the truck.

$s = \frac{150}{t}$

$s = \frac{150}{1.5} = 100$

Speed of the car.

$s + 20$

$= 100 + 20$

$= 120$

Check

$EQ1(\text{car})$ $s + 20 = \frac{180}{t}$

| LHS | RHS |
|---|---|
| $s + 20$ | $\frac{180}{t}$ |
| $= 100 + 20$ | $= \frac{180}{1.5}$ |
| $= 120$ | $= 120$ |

LHS = RHS

$EQ2(\text{truck})$ $s = \frac{150}{t}$

| LHS | RHS |
|---|---|
| $s$ | $\frac{150}{t}$ |
| $= 100$ | $= \frac{150}{1.5}$ |
| | $= 100$ |

LHS = RHS

Close

The speed of the car is 120 $\frac{\text{km}}{\text{hr}}$, and the speed of the truck is 100$\frac{\text{km}}{\text{hr}}$.

10.

Evaluate

| | Case 1 | Case 2 |
|---|---|---|
| Distance | 600 km | 600 km |
| Speed | $s$ | $s$+20 |
| time | $t$ | $t - 1$ |

Interpret

$$\text{speed} = \frac{\text{distance}}{\text{time}}$$

$$s = \frac{d}{t}$$

$EQ1$ $s = \frac{600}{t}$

$EQ2$ $s + 20 = \frac{600}{t-1}$

Solve

Substitute EQ1 into EQ2.

EQ2 $s + 20 = \frac{600}{t-1}$

$(t)(t-1)\left[\left(\frac{600}{t}\right)+20\right]=\left[\frac{600}{t-1}\right](t)(t-1)$

$600(t-1)+20(t^2-t)=600t$

$600t-600+20t^2-20t=600t$

$20t^2-20t-600=0$

$t^2-t-30=0$

$(t-6)(t+5)=0$

$t-6=0$ *or* $t+5=0$
$t=6$ $t=-5$

Since time cannot be negative, $t=6$.

Substitute the value of t, into EQ1.

$s=\frac{600}{t}$

$s=\frac{600}{6}=100$

Check

$EQ1$ $s=\frac{600}{t}$

| LHS | RHS |
|---|---|
| $s$ | $\frac{600}{t}$ |
| $=100$ | $=\frac{600}{6}$ |
| | $=100$ |

LHS = RHS

$EQ2$ $s+2=\frac{600}{t-1}$

| LHS | RHS |
|---|---|
| $s+2$ | $\frac{600}{t-1}$ |
| $=100+20$ | $=\frac{600}{6-1}$ |
| $=120$ | $=\frac{600}{5}$ |
| | $=120$ |

LHS = RHS

Close

The speed of the car is 100 $\frac{\text{km}}{\text{hr}}$.

12.

Evaluate

Joe $\Rightarrow$ 1.5 hours $\Rightarrow a$
Jamil $\Rightarrow$ 2.5 hours $\Rightarrow b$
time together $\Rightarrow t$

Interpret

$\frac{t}{a}+\frac{t}{b}=1$

Solve

$\frac{t}{1.5}+\frac{t}{2.5}=1$

$(1.5)(2.5)\left[\frac{t}{1.5}+\frac{t}{2.5}\right]=\left[1\right](1.5)(2.5)$

$2.5t+1.5t=3.75$

$t=\frac{3.75}{4}=0.9375$

Therefore the time taken to work together is 0.9375 hours. hours.

Check

| LHS | RHS |
|---|---|
| $\frac{t}{a}+\frac{t}{b}$ | 1 |
| $=\frac{0.9375}{1.5}+\frac{0.9375}{2.5}$ | |
| $=0.625+0.375$ | |
| $=1$ | |

LHS=RHS

Close

It would take 0.9375 hours or (0.9375 × 60=56.25 minutes) if the two work together.

14.

Evaluate

Old spray gun $\Rightarrow$ 3 hours $\Rightarrow a$
New spray gun $\Rightarrow$ 1.5 hours $\Rightarrow b$
time together $\Rightarrow t$

Interpret

$$\frac{t}{a}+\frac{t}{b}=1$$

Solve

$$\frac{t}{3}+\frac{t}{1.5}=1$$

$$(3)(1.5)\left[\frac{t}{3}+\frac{t}{1.5}\right]=\left[1\right](3)(1.5)$$

$$1.5t+3t=4.5$$

$$4.5\,t=4.5$$

$$t=1$$

Check

| LHS | RHS |
|---|---|
| $\frac{t}{3}+\frac{t}{1.5}$ | 1 |

$=\frac{1}{3}+\frac{1}{1.5}$

$=\frac{1.5+3}{4.5}$

$=\frac{4.5}{4.5}$

$=1$

LHS=RHS

Close

The time taken to paint the car if the two guns are used together would be 1 hour.

16.

Evaluate

machine 1 $\Rightarrow$ 6 hours $\Rightarrow a$
machine 2 $\Rightarrow$ 4 hours $\Rightarrow b$
time together $\Rightarrow t$

Interpret

$$\frac{t}{a}+\frac{t}{b}=1$$

Solve

$$\frac{t}{6}+\frac{t}{4}=1$$

$$(6)(4)\left[\frac{t}{6}+\frac{t}{4}\right]=\left[1\right](6)(4)$$

$$4t+6t=24$$

$$10\,t=24$$

$$t=2.4$$

Check

| LHS | RHS |
|---|---|
| $\frac{t}{6}+\frac{t}{4}$ | 1 |

$=\frac{2.4}{6}+\frac{2.4}{4}$

$=\frac{9.6+14.4}{24}$

$=\frac{24}{24}$

$=1$

LHS=RHS

Close

The time taken to sort the truckload of refuse if the two machine are used together would be 2.4 hours (2 hours and 24 minutes).

Exercise 13.4

2.

$$\frac{\frac{1}{x}+1}{\frac{1}{x}-2} = \left[\frac{\frac{1}{x}+1}{\frac{1}{x}-2}\right]\left(\frac{x}{x}\right)$$

$$= \frac{\frac{x}{x}+x}{\frac{x}{x}-2x} = \frac{1+x}{1-2x}$$

4.

$$\frac{t+\frac{1}{t}}{-\frac{1}{t}+t} = \left[\frac{t+\frac{1}{t}}{-\frac{1}{t}+t}\right]\left(\frac{t}{t}\right) = \frac{t^2+\frac{t}{t}}{-\frac{t}{t}+t^2}$$

$$= \frac{t^2+1}{-1+t^2} = \frac{t^2+1}{t^2-1}$$

6.

$$\frac{\frac{1}{t}+\frac{2}{x}}{\frac{1}{t}-\frac{3}{x}} = \left[\frac{\frac{1}{t}+\frac{2}{x}}{\frac{1}{t}-\frac{3}{x}}\right]\left(\frac{tx}{tx}\right) = \frac{\frac{tx}{t}+\frac{2tx}{x}}{\frac{tx}{t}-\frac{3tx}{x}}$$

$$= \frac{x+2t}{x-3t}$$

8.

$$\frac{\frac{t^2-u^2}{tu}}{\frac{t+u}{u}} = \frac{t^2-u^2}{tu} \times \frac{u}{t+u} = \frac{t^2-u^2}{t(t+u)}$$

$$= \frac{(t-u)(t+u)}{t(t+u)}$$

$$= \frac{t-u}{t}$$

10.

$$\frac{1+\frac{1}{3t}}{t+\frac{5}{7t}} = \left[\frac{1+\frac{1}{3t}}{t+\frac{5}{7t}}\right]\left(\frac{21t}{21t}\right) = \frac{21t+7}{21t^2+15}$$

12.

$$\frac{\frac{1}{x^2}-\frac{1}{y^2}}{\frac{1}{x^3}+\frac{1}{y^3}} = \left[\frac{\frac{1}{x^2}-\frac{1}{y^2}}{\frac{1}{x^3}+\frac{1}{y^3}}\right]\left(\frac{x^3y^3}{x^3y^3}\right) = \frac{\frac{x^3y^3}{x^2}-\frac{x^3y^3}{y^2}}{\frac{x^3y^3}{x^3}+\frac{x^3y^3}{y^3}}$$

$$= \frac{xy^3-x^3y}{y^3+x^3} = \frac{xy(y^2-x^2)}{y^3+x^3} = \frac{xy(y-x)(y+x)}{(y+x)(y^2-xy+x^2)}$$

$$= \frac{xy(y-x)}{y^2-xy+x^2}$$

14.

$$\frac{\frac{4}{x^2\text{-}25}-\frac{3}{x+5}}{\frac{5}{x^2-25}-\frac{2}{x-5}} = \left[\frac{\frac{4}{x^2\text{-}25}-\frac{3}{x+5}}{\frac{5}{x^2-25}-\frac{2}{x-5}}\right]\left(\frac{(x+5)(x-5)}{(x+5)(x-5)}\right)$$

$$= \left[\frac{\frac{4}{(x\text{-}5)(x+5)}-\frac{3}{x+5}}{\frac{5}{(x-5)(x+5)}-\frac{2}{x-5}}\right]\left(\frac{(x+5)(x-5)}{(x+5)(x-5)}\right)$$

$$= \frac{4-3(x-5)}{5-2(x+5)} = \frac{4-3x+15}{5-2x-10} = \frac{-3x+19}{-2x-5}$$

$$= \frac{-1}{-1} \times \frac{-3x+19}{-2x-5} = \frac{3x-19}{2x+5}$$

16.

$$\frac{\frac{1}{x+3}+\frac{2}{x-5}}{\frac{3}{x-5}+\frac{5}{x+3}}$$

$$= \left[\frac{\frac{1}{x+3}+\frac{2}{x-5}}{\frac{3}{x-5}+\frac{5}{x+3}}\right]\left(\frac{(x+3)(x-5)}{(x+3)(x-5)}\right)$$

$$= \frac{\frac{(x+3)(x-5)}{x+3}+\frac{2(x+3)(x-5)}{x-5}}{\frac{3(x+3)(x-5)}{x-5}+\frac{5(x+3)(x-5)}{x+3}}$$

$$= \frac{(x-5)+2(x+3)}{3(x+3)+5(x-5)}$$

$$=\frac{x-5+2x+6}{3x+9+5x-25}=\frac{3x+1}{8x-16}$$

$$=\frac{3x+1}{8(x-2)}$$

18.

$$\frac{\frac{7}{x-5}-\frac{4}{x^2-25}}{\frac{6}{x^2-25}-\frac{2}{x+5}}$$

$$=\left[\frac{\frac{7}{x-5}-\frac{4}{(x+5)(x-5)}}{\frac{6}{(x+5)(x-5)}-\frac{2}{x+5}}\right]\times\frac{(x+5)(x-5)}{(x+5)(x-5)}$$

$$=\frac{\frac{7(x+5)(x-5)}{x-5}-\frac{4(x+5)(x-5)}{(x+5)(x-5)}}{\frac{6(x+5)(x-5)}{(x+5)(x-5)}-\frac{2(x+5)(x-5)}{x+5}}$$

$$=\frac{7(x+5)-4}{6-2(x-5)}=\frac{7x+35-4}{6-2x+10}=\frac{7x+31}{-2x+16}$$

Summary Test Chapter 13

2.

$x+5=0$
$x=\ -5$

4.

$$\frac{x^2-36}{x^2+4x-12}=\frac{(x+6)(x-6)}{(x+6)(x-2)}=\frac{x-6}{x-2}$$

6.

$$\frac{3x^2-7x-6}{2x^2-4x-6}=\frac{(3x+2)(x-3)}{(2x+2)(x-3)}=\frac{3x+2}{2x+2}$$

8.

$$\frac{x^2-9}{x^2}\times\frac{x^2-25}{x^2+3x-10}=\frac{(x-3)(x+3)}{x^2}\times\frac{(x+5)(x-5)}{(x-2)(x+5)}$$

$$=\frac{(x-3)(x+3)(x-5)}{x^2(x-2)}$$

10.

$$\frac{9x^2}{2y^2}\div\frac{3x}{8y^2}=\frac{9x^2}{2y^2}\times\frac{8y^2}{3x}$$

$$=\frac{(3x)(3x)(4)(2y^2)}{(2y^2)(3x)}=12x$$

12.

$$\frac{x^3-27}{x^3+27}\div\frac{x^2-25}{x^2-3x+9}=\frac{x^3-3^3}{x^3+3^3}\div\frac{x^2-5^2}{x^2-3x+9}$$

$$=\frac{(x-3)(x^2+3x+9)}{(x+3)(x^2-3x+9)}\times\frac{x^2-3x+9}{(x+5)(x-5)}=\frac{(x-3)(x^2+3x+9)}{(x+3)(x+5)(x-5)}$$

14.

$x^2-x-6=(x-3)(x+2)$

$x^2-5x-14=(x-7)(x+2)$

$LCM=(x+2)(x-3)(x-7)$

16.

$$\frac{x^3}{xy^2}+\frac{xy}{xy^2}=\frac{x^3+xy}{xy^2}=\frac{x(x^2+y)}{xy^2}=\frac{x^2+y}{y^2}$$

18.

$$\frac{2x^2+2xy}{x^2-y^2}+\frac{3x-3y}{x-y}$$

Note: $x^2-y^2=(x-y)(x+y)$

$LCM=(x-y)(x+y)$

$$=\frac{(2x^2+2xy)+(3x-3y)(x+y)}{x^2-y^2}$$

$$=\frac{2x^2+2xy+3x^2+3xy-3xy-3y^2}{x^2-y^2}$$

$$=\frac{5x^2+2xy-3y^2}{(x+y)(x-y)}$$

20.

$$\frac{5x^2}{xy^2} - \frac{xy}{xy^2} = \frac{5x^2-xy}{xy^2} = \frac{x(5x-y)}{xy^2} = \frac{5x-y}{y^2}$$

22.

$$\frac{2x}{5} + \frac{4x-6}{10} = 1$$

$$(10)\left[\frac{2x}{5} + \frac{4x-6}{10}\right] = \left[1\right](10)$$

$$4x + 4x - 6 = 10$$

$$8x - 6 = 10$$

$$8x = 16$$

$$x = 2$$

24. $$\frac{1}{x-6} + \frac{9}{x+6} = \frac{18}{x^2-36}$$

$$(x-6)(x+6)\left[\frac{1}{x-6} + \frac{9}{x+6}\right] =$$

$$\left[\frac{18}{x^2-36}\right](x-6)(x+6)$$

$$x + 6 + 9(x-6) = 18$$

$$x + 6 + 9x - 54 = 18$$

$$10x - 66 = 0$$

$$10x = 66$$

$$x = 6.6$$

26.

Evaluate

'a number' $\Rightarrow x$

'seven times it's reciprocal $\Rightarrow 7 \times \frac{1}{x}$

Interpret

$$x + \frac{7}{x} = 8$$

Solve

$$x + \frac{7}{x} = 8$$

$$(x)\left[x + \frac{7}{x}\right] = \left[8\right](x)$$

$$x^2 + 7 = 8x$$

$$x^2 - 8x + 7 = 0$$

$$(x-7)(x-1) = 0$$

thus:

$x - 7 = 0$ or $x - 1 = 0$

$x = 7$ $x = 1$

Check

$x = 7$

| LHS | RHS |
|---|---|
| $x + \frac{7}{x}$ | 8 |
| $= 7 + \frac{7}{7}$ | |
| $= 7 + 1$ | |
| $= 8$ | |

LHS=RHS

$x = 1$

| LHS | RHS |
|---|---|
| $x + \frac{7}{x}$ | 8 |
| $= 1 + \frac{7}{1}$ | |
| $= 1 + 7$ | |
| $= 8$ | |

LHS=RHS

Close

The numbers are 1 and 7.

# CHAPTER 14: Trigonometry

Exercise 14.2

2.a.
$$a^2 + b^2 = c^2$$
$$7^2 + b^2 = 12^2$$
$$49 + b^2 = 144$$
$$b^2 = 144 - 49$$
$$b^2 = 95$$
$$b = \sqrt{95}\text{ cm}$$
$$b \approx 9.75\text{ cm}$$

The length of the second side is approximately 9.75 cm.

b.
$$a^2 + b^2 = c^2$$
$$5^2 + 3^2 = c^2$$
$$25 + 9 = c^2$$
$$c^2 = 34$$
$$c = \sqrt{34}\text{ m}$$
$$c \approx 5.83\text{ m}$$

The length of the hypotenuse is approximately 5.83 m.

c.
$$a^2 + b^2 = c^2$$
$$8^2 + 3^2 = c^2$$
$$64 + 9 = c^2$$
$$c^2 = 73$$
$$c = \sqrt{73}\text{ mm}$$
$$c \approx 8.54\text{ mm}$$

The length of the hypotenuse is approximately 8.54 mm.

d.
$$a^2 + b^2 = c^2$$
$$7^2 + b^2 = 15^2$$
$$49 + b^2 = 225$$
$$b^2 = 225 - 49$$
$$b^2 = 176$$
$$b = \sqrt{176}\text{ mm}$$
$$b \approx 13.27\text{ mm}$$

The length of the second side is approximately 13.27 mm.

e.
$$a^2 + b^2 = c^2$$
$$12^2 + 12^2 = c^2$$
$$144 + 144 = c^2$$
$$c^2 = 288$$
$$c = \sqrt{288}\text{ m}$$
$$c \approx 16.97\text{ m}$$

The length of the hypotenuse is approximately 16.97 m.

f.
$$a^2 + b^2 = c^2$$
$$5^2 + 5^2 = c^2$$
$$25 + 25 = c^2$$
$$c^2 = 50$$
$$c = \sqrt{50}\text{ m}$$
$$c \approx 7.07\text{ m}$$

The length of the hypotenuse is approximately 7.07 m.

g. Convert all sides to the same unit of measurement: 1m = 100cm

$$a^2 + b^2 = c^2$$
$$100^2 + 65^2 = c^2$$
$$10000 + 4225 = c^2$$
$$c^2 = 14225$$
$$c = \sqrt{14225}\text{ cm}$$
$$c \approx 119.27\text{ cm}$$

The length of the hypotenuse is approximately 119.27 cm or 1.1927 m.

h.
$$a^2 + b^2 = c^2$$
$$2^2 + b^2 = (\sqrt{7})^2$$
$$4 + b^2 = 7$$
$$b^2 = 7 - 4$$
$$b^2 = 3$$
$$b = \sqrt{3}\text{ cm}$$
$$b \approx 1.73\text{ cm}$$

The length of the second side is approximately 1.73 cm.

i.
$$a^2 + b^2 = c^2$$
$$\left(\sqrt{3}\right)^2 + b^2 = 15^2$$
$$3 + b^2 = 225$$
$$b^2 = 225 - 3$$
$$b^2 = 222$$
$$b = \sqrt{222}\text{ cm}$$
$$b \approx 14.9\text{ cm}$$

The length of the second side is approximately 14.9 cm.

4. Step1:
Find the vertical height of the triangle:

$$a^2 + b^2 = c^2$$
$$18^2 + b^2 = 25^2$$
$$324 + b^2 = 625$$
$$b^2 = 625 - 324$$
$$b^2 = 301$$
$$b = \sqrt{301}\text{ m}$$
$$b \approx 17.4\text{ m}$$

Step2:
Find the area of the triangle:

Area = $\frac{1}{2}$ base $\times$ height

$$= \frac{1}{2} \times 18 \times 17.4\text{ m}^2$$
$$= 156.6\text{ m}^2$$

6.a. Length of radius = 1 m

Length of each side of the square
= 2 × 1 m
= 2 m

Area of the square = 2 m × 2 m
= 4 m$^2$

b. Length of the radius = 1 m

Therefore, the length of the diagonal in the square = 2 × 1 m
= 2 m

The sides of the square are equal in length, so

$$a^2 + a^2 = c^2$$
$$a^2 + a^2 = 2^2$$
$$2a^2 = 4$$
$$a^2 = 2$$
$$a = \sqrt{2}\text{ m}$$

The sides of the square are $\sqrt{2}$ m.

The area of the square = side$^2$
$= \left(\sqrt{2}\text{ m}\right)^2$
$= 2\text{ m}^2$

The area of the largest square is 2 m$^2$

8. Length of the radius = 2 cm

Therefore, the length of the diagonal in the square = 2 × 2 cm
= 4 cm

The sides of the square are equal in length, so

$$a^2 + a^2 = c^2$$
$$a^2 + a^2 = 2^2$$
$$2a^2 = 16$$
$$a^2 = 8$$
$$a = \sqrt{8}\text{ cm}$$

The sides of the square are $\sqrt{8}$ cm.

The area of the square = side$^2$
$= \left(\sqrt{8}\text{ cm}\right)^2$
$= 8\text{ cm}^2$

The area of the largest square is 8 cm$^2$

10. $a^2 + b^2 = c^2$
$a^2 + 1^2 = 2500^2$
$a^2 = 2499$
$a = \sqrt{2499}$
$a \approx 49.99$ m

The vertical height of the tower is approximately 49.99 m

Exercise 14.3

2.a $\text{Cos A} = \frac{10}{15}$
$\text{Cos A} = 0.6667$
$A = 48.19°$

$\text{Sin B} = \frac{10}{15}$
$\text{Sin B} = 0.6667$
$B = 41.81°$

$\text{Tan A} = \frac{a}{10}$
$\text{Tan } 48.19° = \frac{a}{10}$
$1.1180 = \frac{a}{10}$
$a = 1.1180 \times 10$
$a = 11.18$ cm

b. $\text{Tan } 35 = \frac{a}{10}$
$0.7002 = \frac{a}{10}$
$a = 7.002$ cm

$\text{Cos } 35 = \frac{10}{c}$
$0.8192 = \frac{10}{c}$
$c \times 0.8192 = 10$
$c = \frac{10}{0.8192}$
$c = 12.2077$ cm

$B + 35 + 90 = 180$
$B + 125 = 180$
$B = 55°$

4. Let the angel at the ground = A

$\text{Tan A} = \frac{1.8}{2}$
$\text{Tan A} = 0.9$
$A = 41.99°$
$\text{Tan A} = \frac{h}{6}$
$\text{Tan } 41.99° = \frac{h}{6}$
$0.9 = \frac{h}{6}$
$h = 5.4$ m

( Note that this could have been calculated using similar triangles.)

Exercise 14.4

2. Let the distance that the balloon travelled = a
Let the height of the balloon = b = 200 m

A = change in angel of elevation
= 90° - 75°
= 15°

$\text{Tan A} = \frac{a}{b}$
$\text{Tan } 15 = \frac{a}{200}$
$0.2679 = \frac{a}{200}$
$a = 53.59$ m

The balloon had travelled 53.59 m in 1 minute

In 60 minutes (1 hour), the balloon would travel:

$60 \times 53.59 \text{ m} = 3215.4 \text{ m}$
$= 3.2154 \text{ km}$

$\text{Average speed} = \frac{\text{distance}}{\text{time}}$
$= \frac{3.2154 \text{ km}}{1 \text{ hour}}$
$= 3.2154 \frac{\text{km}}{\text{hr}}$

The average speed of the balloon is approximately 3.22 $\frac{\text{km}}{\text{hr}}$

4. Draw a diagram of the slide, using a right -angled triangle with hypotenuse length = 10m, and vertical height = 8m. Let the angle of the slide = A

$$\text{Sin A} = \frac{8}{10}$$
$$\text{Sin A} = 0.8$$
$$\text{A} = 53.13°$$

The angle at the steepest part of the slide is 53.13°. Yes, you can host the Championship.

6. See the diagram for all 3 parts.

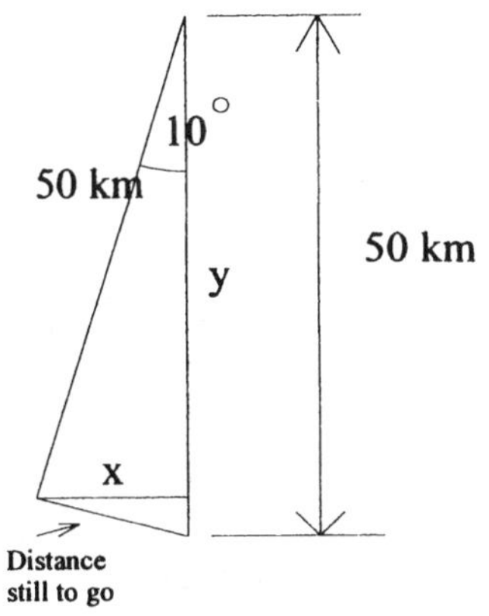


a. Let $x$ be the distance that they are west of the line that they should have been travelling.

$$\text{Sin } 10 = \frac{x}{50}$$
$$0.1736 = \frac{x}{50}$$
$$x = 8.68\,\text{km}$$

They are approximately 8.68 kilometres west of where they should be.

b. Let $y$ be the distance that they have travelled in the direction of south.

$$\cos 10 = \frac{y}{50}$$
$$0.9848 = \frac{y}{50}$$
$$y = 49.24\,\text{km}$$

They have travelled 49.24 kilometres in a southerly direction, so they are still:

50 km – 49.42 km
= 0.76 km north of Lynn Lake.

c. Use Pythagoras to calculate the length of the hypotenuse in the smaller triangle.

$$c^2 = a^2 + b^2$$
$$c^2 = 0.76^2 + 8.68^2$$
$$c^2 = 0.5776 + 75.3424$$
$$c^2 = 75.92$$
$$c = \sqrt{75.92}$$
$$c \approx 8.71$$

They still need to travel 8.71 km to reach Lynn Lake.

Exercise 14.7.2

2.

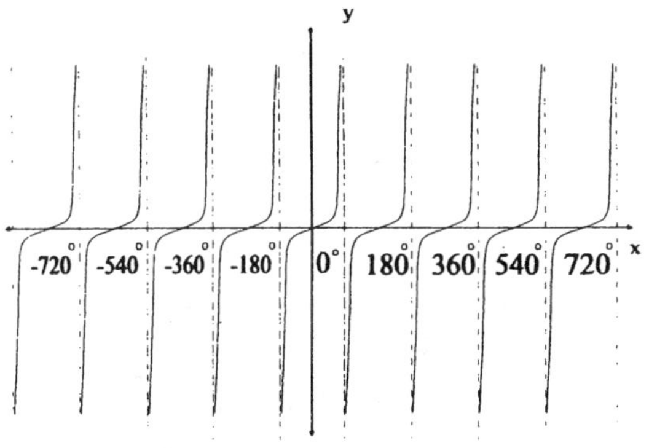


Exercise 14.7.3

2.a.

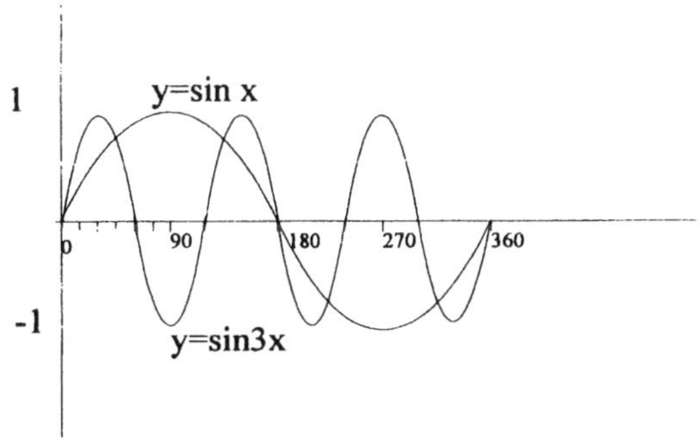

b.

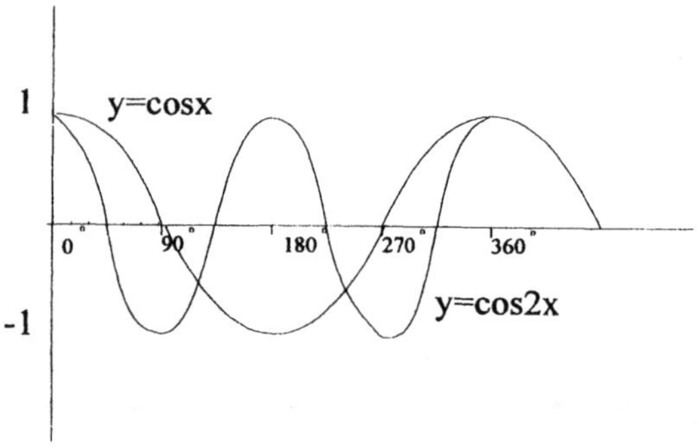


c. If $a$ is greater than 1, then it increases the number of cycles of the curve; if $a$ is less than 1, then it decreases the number of cycles.

d. The $a$ will have the same effect as it does on the Sin graph.

e. The $a$ will alter the number of complete cycles.

f. Complete cycle of curve every $\frac{180°}{a}$ for the Tan graph, and $\frac{360°}{a}$ for Sin and Cos.

4. Maximum value = + 2, minimum value = – 2. That is, a complete cycle is still from 0° to 360°, but the height of the curve is increased.

Exercise 14.8

2.a.

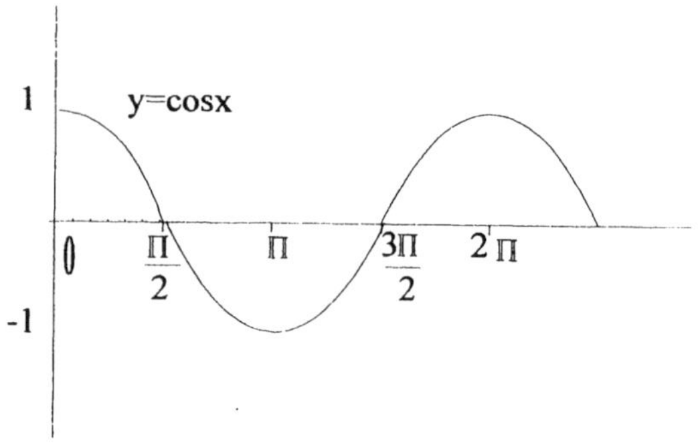


b.

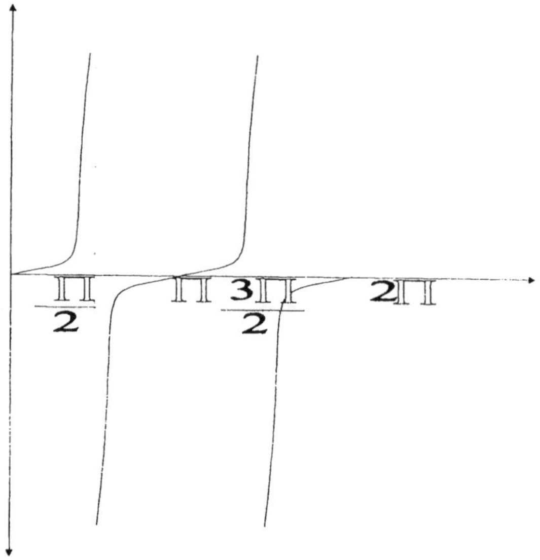


Exercise 14.9.1

2. Test the triangle with the Sine Rule:

$$\frac{a}{\text{Sin A}} = \frac{12}{\text{Sin }55} = \frac{12}{0.8192} = 14.6484$$

$$\frac{b}{\text{Sin B}} = \frac{11}{\text{Sin }62} = \frac{11}{0.8829} = 12.4589 \neq \frac{a}{\text{Sin A}}$$

Because $\frac{a}{\text{Sin A}} = \frac{b}{\text{Sin B}}$ FOR ANY TRIANGLE, then this shape cannot exist.

4. Angle at course correction
= 45° + 90° + 30°
= 165°

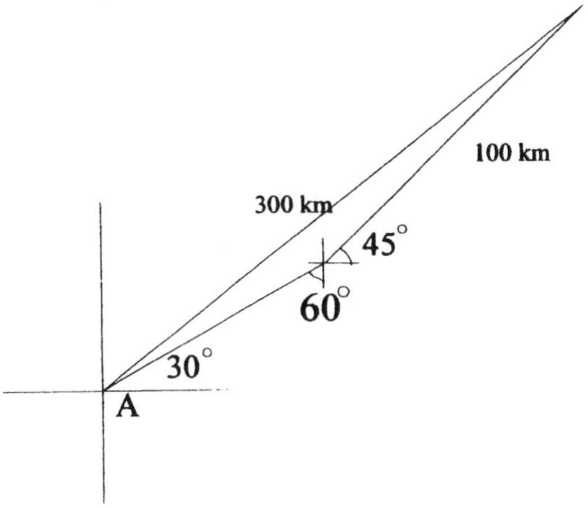


$\frac{a}{\sin A} = \frac{b}{\sin B}$

$\frac{100}{\sin A} = \frac{300}{\sin 165}$

$\frac{100}{\sin A} = \frac{300}{0.2588}$

$\frac{100}{\sin A} = 1159.1963$

$100 = 1159.1963 \times \sin A$

$\sin A = \frac{100}{1159.1963}$

$\sin A = 0.0863$

$A = 4.95°$

( Note that this angle cannot be obtuse because another angle is obtuse).

The angle at A is approximately 5°.

The bearing of Ucluelet is:
$90° - 5° - 30° = $ N55°E

Exercise 14.9

2. Let the width of the swamp be $X$.

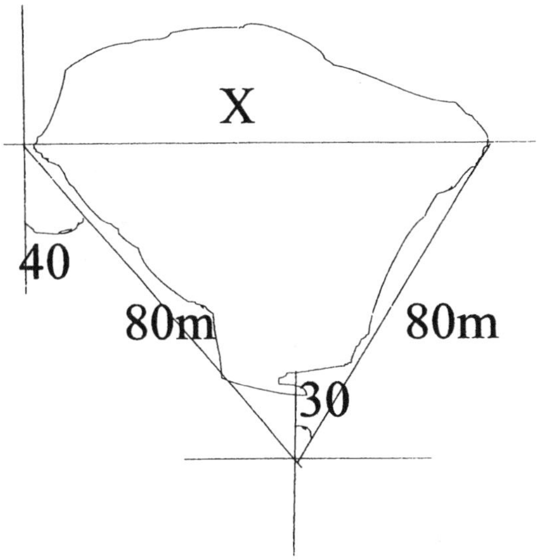


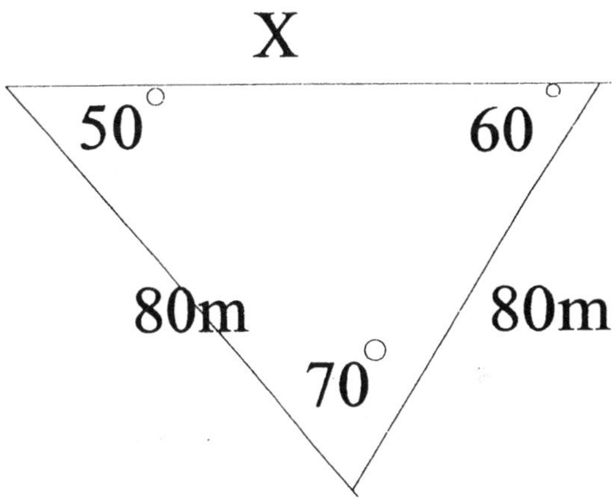


This is a Cosine Rule problem because two side lengths are know, and all angles are known.

$c^2 = a^2 + b^2 - 2ab \operatorname{Cos} C$

$X^2 = 80^2 + 80^2 - 2 \times 80 \times 80 \operatorname{Cos} 70$

$= 6400 + 6400 - 2 \times 6400 \times 0.342$
$= 12800 - 12800 \times 0.342$
$= 12800 - 4377.6$
$= 8422.4$

$X \approx 91.77$ m

The width of the swamp is approximately 91.77 m.

4. Let the height of the tower be H.

Step 1: Find the length of the hypotenuse of the large right-angled triangle by using the Sine Rule with the smaller, right-hand triangle.

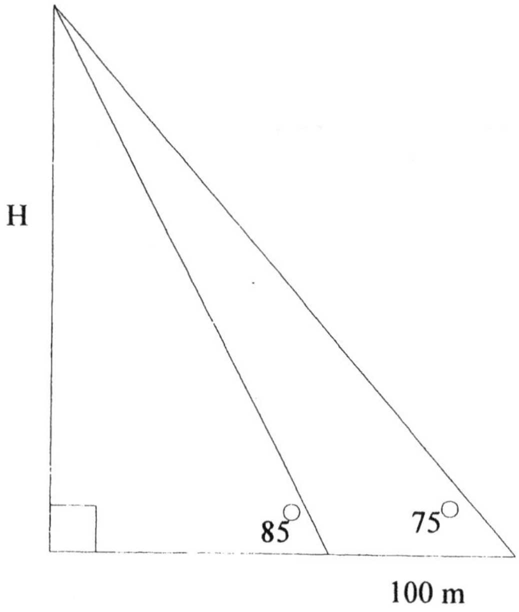


One angle is given as 75°. Another angle is 180° − 85° = 95°. This means that the third angle is 10°.

$\frac{c}{\sin 95} = \frac{100}{\sin 10}$

$\frac{c}{0.9962} = \frac{100}{0.1736}$

$c = 0.9962 \times 576.03687$

$c = 573.84793$

Step 2: Use the Sine ratio for the larger right angled triangle, with hypotenuse of length 573.84793, and the angle of 75°.

$\text{Sin } 75 = \frac{H}{573.84793}$

$0.9659 = \frac{H}{573.84793}$

$H = 0.9659 \times 573.84793$

$H = 554.29$

The height of the CN Tower is approximately 554 m.

6. Test this triangle with the Cosine Rule:

$c^2 = a^2 + b^2 - 2ab \cos C$

Left Hand Side: $20^2 = 400$

Right Hand Side:

$12^2 + 10^2 - 2 \times 12 \times 10 \times \cos 160$

$= 144 + 100 - 240 \times \cos 160$

$= 244 - 240 \times (-0.9397)$

$= 244 + 225.528$

$= 469.528$

The Left Hand Side does not equal the Right Hand Side, and so the Cosine Rule has shown that this "triangle" cannot exist.

Exercise 14.10

2. a. 0.45295

b. 0.54097

c. 572.95721

d. 0.82871

e. 0.86949

f. 0.56212

## End of Chapter Test

2. Let the length of AB = $x$

$\frac{x}{12} = \sin 25$

$\frac{x}{12} = 0.4226$

$x = 12 \times 0.4226$

$x = 5.0712$

Since $\angle B = 90°$, then $\angle A = 65°$.

Let the length of CB = $y$

$\frac{y}{12} = \cos 25$

$\frac{y}{12} = 0.9063$

$y = 12 \times 0.9063$

$y = 10.8765$

The angles and lengths (approximate) are 90°, 65°, 10.88 cm, 5.07 cm.

4.a. $60° = \pi \times \frac{60}{180}$

$= \frac{\pi}{3}$ radians

b. $120° = \pi \times \frac{120}{180}$

$= \frac{2\pi}{3}$ radians

c. $64° = \pi \times \frac{64}{180}$

$= \pi \times \frac{32}{90}$

$= \frac{16\pi}{45}$ radians

d. $320° = \pi \times \frac{320}{180}$

$= \pi \times \frac{160}{90}$

$= \frac{16\pi}{9}$ radians

6.

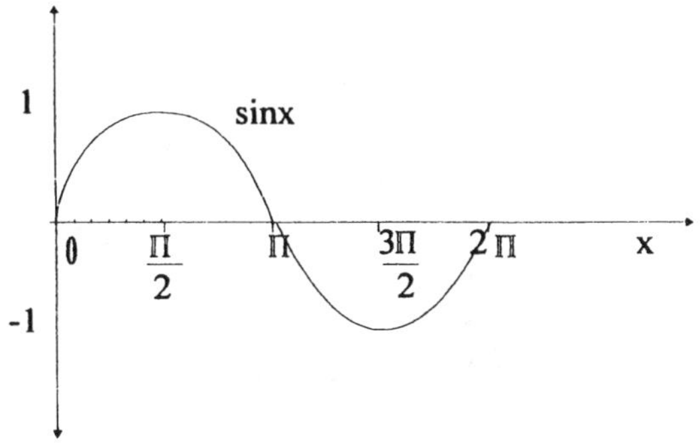


8.

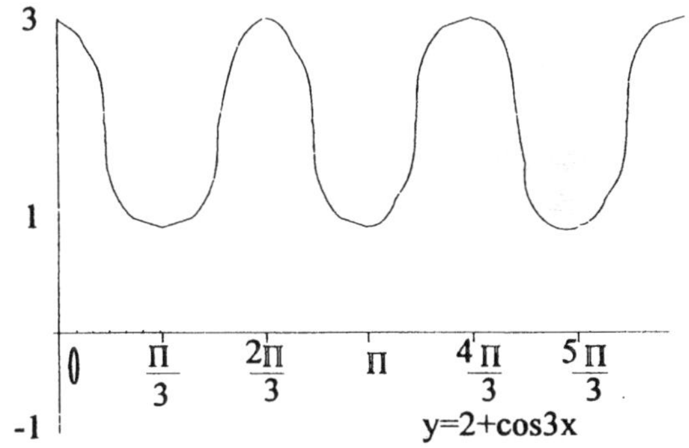


10. Let the width of the lake be $x$.

The 3rd angle of the triangle is :
$180° - 120° - 20° = 40°$

Use Sine Rule to find the value of $x$.

$\frac{x}{\sin 20} = \frac{100}{\sin 40}$

$\frac{x}{0.3420} = \frac{100}{0.6428}$

$x = 0.3420 \times 155.56938$

$x = 53.2$

The width of the lake is approximately 53.2 m.

12. Let the distance of the plane from Nipawin be $x$.

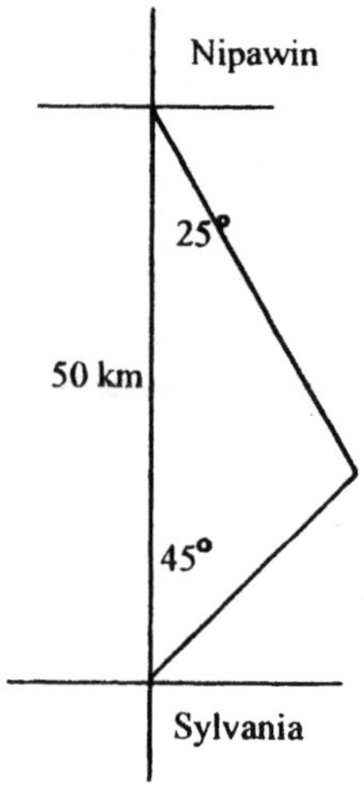


The third angle of the triangle :
$= 180° - 25° - 45°$
$= 110°$

$$\frac{x}{\sin 45} = \frac{50}{\sin 110}$$

$$\frac{x}{0.7071} = \frac{50}{0.9397}$$

$$\frac{x}{0.7071} = 53.208471$$

$$x = 0.7071 \times 53.208471$$

$$x = 37.62$$

The plane is approximately 37.62 km from Nipawin.

# CHAPTER 15: Probability and Statistics

## Exercise 15.2

2. There are approximately 26 weeks from the beginning of May to the end of October. This means that over the next 20 years, you will mow the lawn $20 \times 26 = 520$ times.

$P(\text{injury}) = \frac{1}{3623}$

Over the next 20 years, the number of injuries that you can expect

$= \frac{1}{3623} \times 520 = 0.14$

4. The ratio of wins to losses = 4:5

$P(\text{win}) = \frac{4}{(5+4)} = \frac{4}{9}$

6. $P(\text{quadruplets}) = \frac{1}{900000}$

Expected number of quadruplets in 10 000 000 births $= \frac{1}{900000} \times 10000000$

$= \frac{100}{9}$

$= 11.1$

That is, you can expect 11 sets of quadruplets in 10 000 000 births.

## Exercise 15.3

2.

| Die / Coin | 1 | 2 | 3 | 4 | 5 | 6 |
|---|---|---|---|---|---|---|
| H | H1 | H2 | H3 | H4 | H5 | H6 |
| T | T1 | T2 | T3 | T4 | T5 | T6 |

There are 12 possible outcomes, and three of these satisfy the condition that the coin be a Head, and the die be showing an even number.

$P(\text{Head and even number}) = \frac{3}{12} = \frac{1}{4}$

Exercise 15.5

2.a.

BBB, BBG, BGB,BGG,GBB, GBG, GGB,GGG

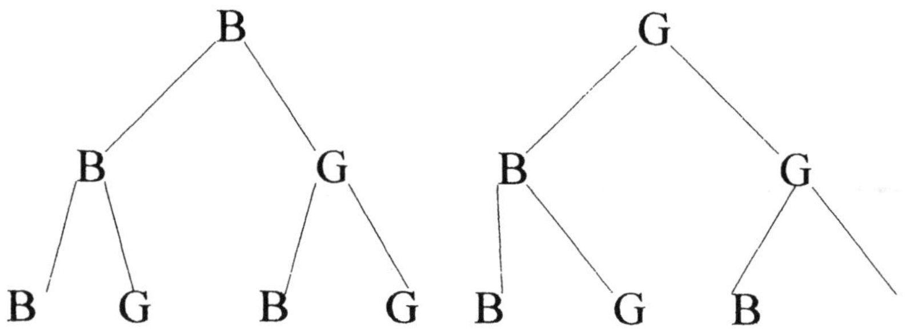


b.

BBBB, BBBG, BBGB, BBGG, BGBB, BGBG, BGGB, BGGG, GBBB, GBBG, GBGB, GBGG, GGBB, GGBG, GGGB, GGGG

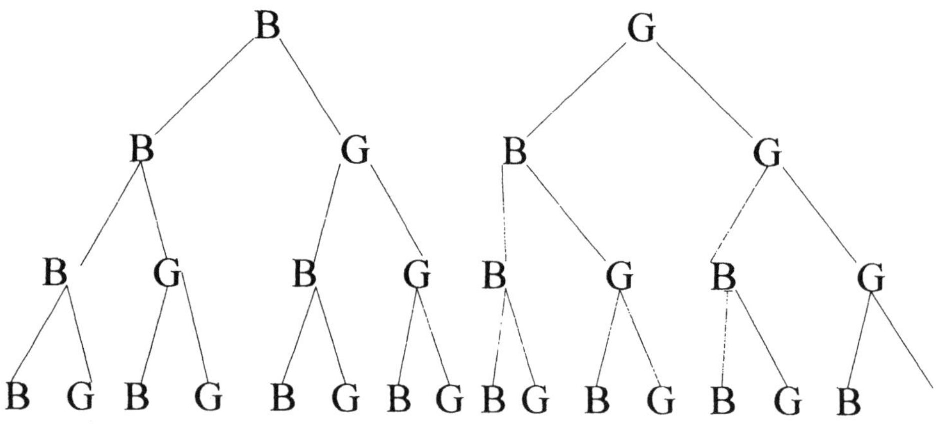


c.

There are 32 possible order/arrangements for 5 children:

BBBBB, BBBGB, BBGBB, BBGGB, BGBBB, BGBGB, BGGBB, BGGGB, GBBBB, GBBGB, GBGBB, GBGGB, GGBBB, GGBGB, GGGBB, GGGGB, BBBBG, BBBGG, BBGBG, BBGGG, BGBBG, BGBGG, BGGBG, BGGGG, GBBBG, GBBGG, GBGBG, GBGGG, GGBBG, GGBGG, GGGBG, GGGGG

There are 10 outcomes with only two girls:
BBGGB, BGBGB, BGGBB, GBBGB, GBGBB, GGBBB, BBBGG, BBGBG, BGBBG, GBBBG

P(2 girls in 5 children) $= \frac{10}{32} = \frac{5}{16}$

Exercise 15.7

2. Depends upon time of year. Find the maximin temperature for all 7 days last week, add them, and then divide by 7.

4.b. Average

$$= \frac{25+30+50+45+32+42+46+38+42+31}{10}$$

$$= \frac{381}{10}$$

$$= 38.1$$

Exercise 15.8

2. Average budget before the cuts:

$$\frac{\$\ 50\,000+80\,000+60\,000+40\,000+75\,000}{5}$$

$$= \frac{\$\ 305\,000}{5}$$

$$= \$61\,000$$

Marketing's new budget = 75% of the of budget

$$= 75\% \times \$80\,000$$

$$= \$60\,000$$

New average budget =

$$\frac{\$\ 50\,000+60\,000+60\,000+40\,000+75\,000}{5}$$

$$= \frac{\$285\,000}{5}$$

$$= \$57\,000$$

Marketing is still above average; the vice-president was correct.

Exercise 15.9

2. Average time

$= \frac{500+520+600+480+470+540+550+600+602+580+575+800+490+532+603+227+353+619+482+532}{20}$

$= \frac{10655}{20}$

$= 532.75$ hours

Median time:

227, 353, 470, 480, 482, 490, 500, 520, 532, 532, 540, 550, 575, 580, 600, 600, 602, 603, 800, 619

Because there is an even number of scores, we will need to average the two "middle" scores

10th score = 532 hours
11th score = 540 hours

Median $= \frac{532+540}{2}$ hours

$= \frac{1072}{2}$ hours

= 536 hours

Exercise 15.10

2.a. note that the probability of throwing any number is $\frac{1}{6}$.

In 60 throws, you *expect* that each face will occur 10 times.

| Outcomes: | 1 | 2 | 3 | 4 | 5 | 6 |
|---|---|---|---|---|---|---|
| Observed frequency | 8 | 11 | 12 | 7 | 11 | 11 |
| Expected frequency | 10 | 10 | 10 | 10 | 10 | 10 |
| Expected - Observed | 2 | -1 | -2 | 3 | -1 | -1 |
| (Expected - Observed)$^2$ | 4 | 1 | 4 | 9 | 1 | 1 |
| $\frac{(\text{Expected - Observed})^2}{\text{Expected}}$ | 0.4 | 0.1 | 0.4 | 0.9 | 0.1 | 0.1 |
| | $\chi^2 = 0.4 + 0.1 + 0.4 + 0.9 + 0.1 + 0.1 = 2.0$ | | | | | |

There are 5 degrees of freedom (because there are 6 faces on the die), and so the critical value of $\chi^2 = 11.07$.
You have to conclude that the die is fair because you have not exceeded the critical value.

b.

| Outcomes: | 1 | 2 | 3 | 4 | 5 | 6 |
|---|---|---|---|---|---|---|
| Observed frequency | 80 | 110 | 120 | 70 | 110 | 110 |
| Expected frequency | 100 | 100 | 100 | 100 | 100 | 100 |
| Expected - Observed | 20 | -10 | -20 | 30 | -10 | -10 |
| (Expected - Observed)$^2$ | 400 | 100 | 400 | 900 | 100 | 100 |
| $\frac{(\text{Expected - Observed})^2}{\text{Expected}}$ | 4 | 1 | 4 | 9 | 1 | 1 |
| | $\chi^2 = 4 + 1 + 4 + 9 + 1 + 1 = 20$ | | | | | |

There are 5 degrees of freedom (because there are 6 faces on the die), and so the critical value of $\chi^2 = 11.07$.

You have to conclude that the die is *not* fair because you have exceeded the critical value.

Exercise 15.11

2. Total energy requirements
= (2 000 000+5 000 000+1 500 000+2 000 000+1 500 000) joules
= 12 000 000 joules

| Activity | joules | Portion of the total | Portion as fraction | Angle |
|---|---|---|---|---|
| Sleeping | 2 000 000 | $\frac{2\,000\,000}{12\,000\,000}$ | $\frac{1}{6}$ | $\frac{1}{6} \times 360° = 60°$ |
| Moderate work | 5 000 000 | $\frac{5\,000\,000}{12\,000\,000}$ | $\frac{5}{12}$ | $\frac{5}{12} \times 360° = 150°$ |
| Reading | 1 500 000 | $\frac{1\,500\,000}{12\,000\,000}$ | $\frac{3}{24}$ | $\frac{3}{24} \times 360° = 45°$ |
| Exercise | 2 000 000 | $\frac{2\,000\,000}{12\,000\,000}$ | $\frac{1}{6}$ | $\frac{1}{6} \times 360° = 60°$ |
| Other | 1 500 000 | $\frac{1\,500\,000}{12\,000\,000}$ | $\frac{3}{24}$ | $\frac{3}{24} \times 360° = 45°$ |

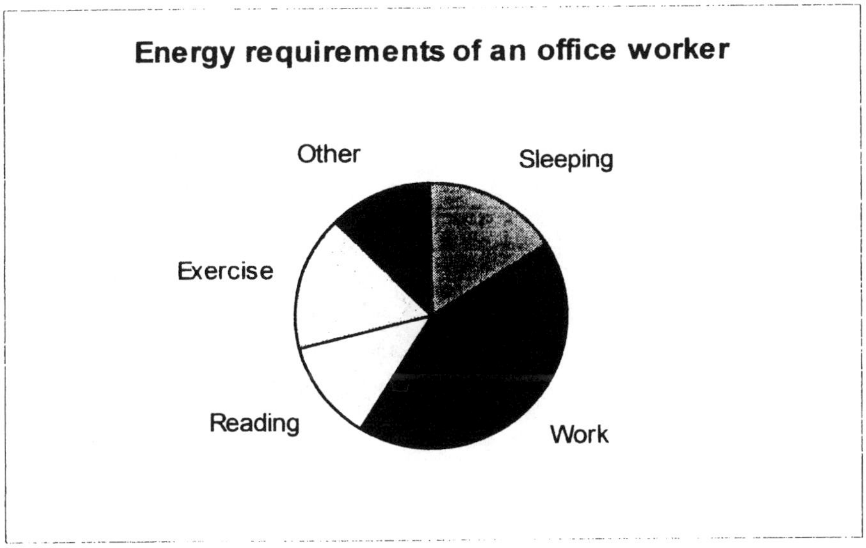

4.a.

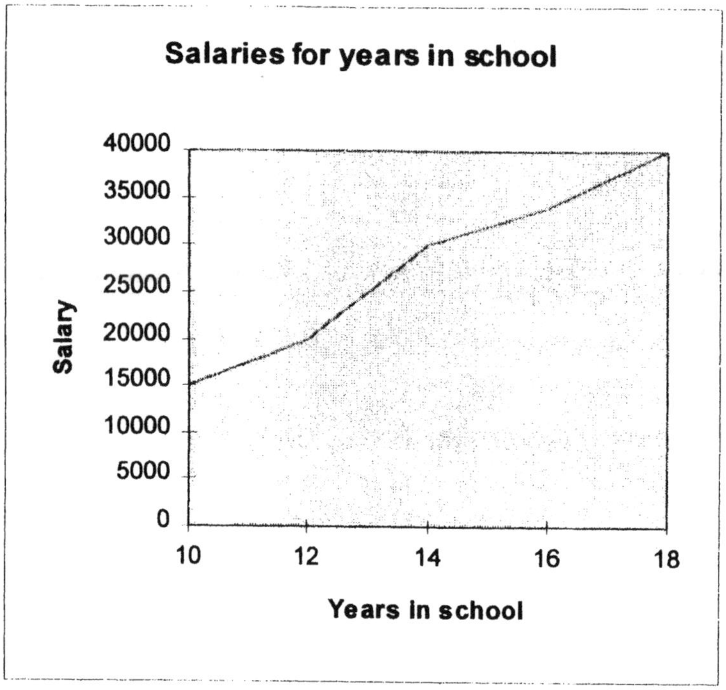


There does appear to be a relationship between number of years in school and the starting salary for a job - the more years in school, the higher the starting salary.

b. A line graph is appropriate because:

i. we are looking at information over time;
ii. there is no "total" amount of salary to which we can compare individual amounts; and
iii. Although we are not told about values for all possible times (such as 15 years), we can assume that data does exist, and also fits this pattern.

Exercise 15.12

2.

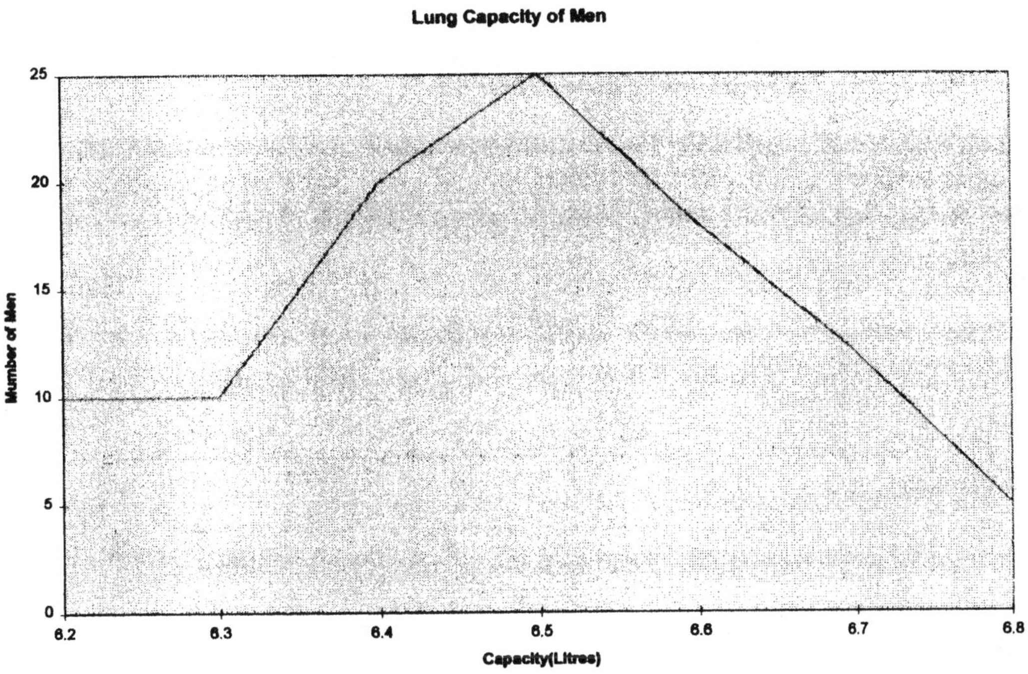


The shape of the graph is similar to a normal curve We would expect the shape of the graph to be normal with a sample of 100 people. The graph would be even more like a normal curve is we measured the volume of the lungs with greater accuracy (say to 3 decimal places).

4.

| Cause of death | Probability |
|---|---|
| Motor vehicle accident | $\frac{1}{150}$ |
| Bicycle accident | $\frac{1}{25\,000}$ |
| plane crash | $\frac{1}{30\,000}$ |
| Bus accident | $\frac{1}{500\,000\,000}$ |

Assume the that there have been 1 000 000 000 accidents. Then, the numbers of people killed in each type of transportation accident are:

| Cause of death | Number of people killed: |
|---|---|
| Motor vehicle accident | 6666667 |
| Bicycle accident | 40000 |
| Plane crash | 33333 |
| Bus accident | 2 |

A bar graph should be used to display this information because we have no information about the other causes of death. We could use a pie chart, and label a segment as "Other Causes", but, because we are only interested in trasportation-related deaths, we do not need to do that.

Note that the very low number for bus accidents, and the very high number of car crashes, causes a problem with the graph, and it requires a very large sheet of paper to draw.

End of Chapter Test

2. There are 52 cards in a regular deck, and two of these are black jacks.

$$P(\text{Black Jack}) = \frac{2}{52}$$

$$= \frac{1}{26}$$

You would expect to need to be dealt 26 cards before you would see a black jack.

4. If there was no difference between the colas, then you would *expect* that the same number of people would select each cola. This is a chi-square problem.

| Cola: | A | B | C | D |
|---|---|---|---|---|
| Number who selected this one as best | 220 | 200 | 185 | 195 |
| Expected number | 200 | 200 | 200 | 200 |
| Expected - Actual | -20 | 0 | 15 | 5 |
| (Expected -Actual)$^2$ | 400 | 0 | 225 | 25 |
| $\frac{\text{(Expected - Actual)2}}{\text{Expected}}$ | 2 | 0 | 1.125 | 0.0125 |
| | $\chi^2 = 2 + 0 + 1.125 + 0.125 = 3.25$ | | | |

Because there are 4 types of colas being tested, we look up the critical value of $\chi^2$ with 3 degrees of freedom; this is 7.82. The value that we have calculated for $\chi^2$ does not exceed this value, and so there is no difference between the colas in terms of preference.

6. The average price of the houses is \$241 000, while the median price is \$290 000. All but one of the house prices are at least \$250 000, and three houses are greater than \$300 000. But, the 1 house with the low price reduces the average by a large amount.

In general, averages are more susceptible to being influenced by extremely high or low scores.

8.a.

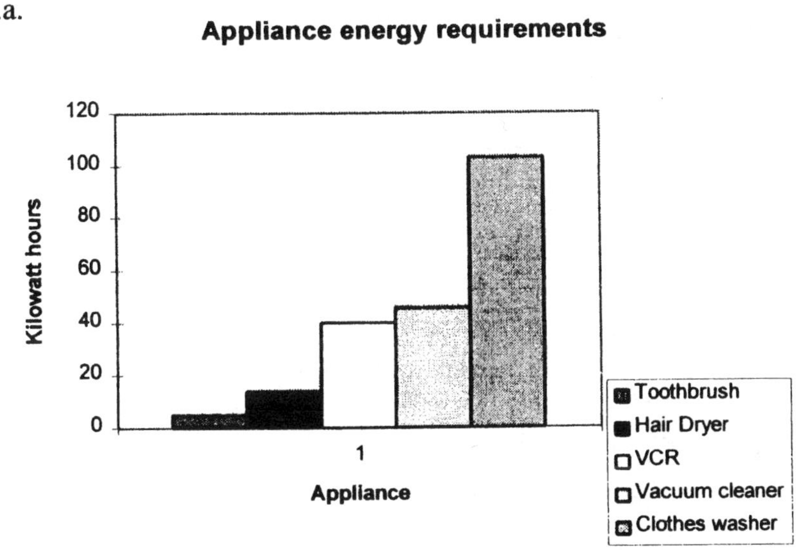


b. No, because the appliance listed do not represent a complete list of home appliance. Those listed are given to illustrate relative energy requirement, not to account for all energy usage.